Meltdown

RICHARD THOMPSON

ISBN 978-1-63784-723-7 (paperback)
ISBN 978-1-63784-724-4 (digital)

Hawes & Jenkins Publishing
16427 N Scottsdale Road Suite 410
Scottsdale, AZ 85254
www.hawesjenkins.com

Printed in the United States of America

Chapter 1

Antarctica is the fifth-largest continent and the world's highest continent. The climate of Antarctica is characterized by extreme cold and high winds. According to some scientists estimations, the conversion of all of Antarctica's ice water would raise global sea levels by fifty to sixty-five meters. Antarctica's ice sheets contain seventy percent of the world's fresh water in a frozen state, and because the cold air holds water, there is little water for vapor precipitation.

Antarctica is also the driest continent on earth. Because of these harsh conditions, it was always hypothesized that Antarctica had no indigenous people. The only humans thought to have lived there were scientists at research stations.

In 1973, Dr. Elizabeth Fehlinger, a distinguished archaeologist, and her six-member team made a miraculous find in the pristine wilderness of the most remote region in Antarctica that would change the hypothesis forever.

Beneath the earth's frozen surface in an icy cavern was what Dr. Fehlinger believed to be possibly one of the first indigenous people of the Antarctic. It took two years of painfully slow work using very specialized tools and equipment to evacuate her discovery. To maintain the integrity of the specimen, he or she was left embedded inside the block of ice that was the find's final resting place. The edges of the block of ice were carefully chipped away until they gave way to gravity. Then, the block of ice was placed right inside a large, specially designed refrigerator that mimicked the temperature of the icy cavern.

Dr. Fehlinger, her six-member team, and the refrigerator were all flown on a United States military cargo plane to Anchorage, Alaska. The refrigerator was then unloaded onto a truck and driven to the prestigious University of Alaska Museum, or UAM for short. The specimen was placed inside a temperature-controlled storage room for examination and study under the careful eye of Dr. Fehlinger.

After six months, tragedy struck Dr. Fehlinger and her six-member team. One by one, they all died a very horrible death from a mysterious illness that turned their skin grayish, stiff, and hard. Their internal organs were almost like stones. Their transmogrification was extremely painful, as they could feel their organs shutting down and hardening. Dr. Fehlinger was the first to succumb to the illness. She never got a chance to finish writing her paper on what she called the Antarctic's first inhabitants.

Between the years of 1973, when Dr. Fehlinger first made her discovery, and 1983, there was a noticeable shift in the earth's climate. The summer months were hotter and lasted longer. The winter months were colder and shorter. In between the seasons were never before rain storms, snow, hurricanes, tornadoes, and rising sea levels.

Scientists from all over the world, with the backing and resources of their respective governments', studied this new phenomenon and concluded that industrialization mainly from the western world with its' human activities, principally the injection of carbon dioxide into the atmosphere through the burning of fossil fuels, was warming the earth. This new phenomenon is coined global warming.

Scientists explained that global warming occurs because of an imbalance of the earth's energy. The earth receives solar energy at short wave lengths and reflects energy back into space at higher infrared wave lengths. Gases in the atmosphere have molecules that absorb energy reflected at these longer wave lengths, trapping heat in the earth's atmosphere. For the layperson, scientists refer to these gases as greenhouse gases or the greenhouse effect. The analogy of the green house is used because, as in the atmosphere, the sun's rays enter a greenhouse, warming the plants inside, while the glass absorbs and remits the infrared radiation from soils and plants back into the greenhouse, keeping it warm. There were debates raging during those years between scientists and business conglomerates about whether the observed warming of the earth's surface was due primarily to greenhouse gases generated by human activities or whether the warming was part of a natural variation of the earth's temperature. There were some, although a very small group, who had one thing in common. They were business owners of large industrialized plants and refused to consider the scientific data put forward by the scientist that supported their position. In 2016, the Presidency of the United States was won by one of those like-minded small groups of people. Any funding for further research to resolve the issue and its long-term effects on the climate was cut. Dr. Fehlinger's paper and miraculous discovery were relegated to a corner of a storage unit until, ultimately, they were forgotten.

The summer of 2017 was on track to becoming the hottest summer ever recorded for the State of Alaska since weather statistics began being collected in 1959, when Alaska became the forty-ninth state of the United States that same year.

John had a nagging cough for two weeks. At first, his parents, John Wainwright senior, whom everyone called JW, and his wife Mary thought it was just a passing summer cold or an allergy of some sort that children oftentimes get.

John was eleven and a student at Matthew Alexander Henson elementary school, named after the Black American explorer who accompanied Robert Edwin Peary on seven Arctic expeditions, including the 1909 expedition, where they were the first recorded human beings to reach the north pole. Robert Peary, who was white, was the leader of the expedition. He suffered frost bite, and it was actually Matthew Henson who first set foot on the north pole, but because he was black, he was cheated out of his rightful place in America's history books, and the honor and distinction were given to Robert Peary.

John awoke screaming and in pain in the middle of the night. Mary heard the screams awaking instantly, like only a mother could hear the distressing sounds of her child. She raced into her son's bedroom, flicking on the light. One look at John, and she knew something was terribly wrong. His skin looked grayish, and when she touched his forehead to check his temperature, his skin felt stiff and hard. "John, what's wrong?" She asked with deep concern.

John didn't answer. He looked like he was trying to speak, but couldn't. He looked at her with flat black eyes.

Mary raced back into the bedroom she shared with her husband and shook him by the shoulder frantically. "JW! Wake Up! Something is wrong with John!"

"What do you mean something is wrong with John?" JW asked groggily.

"He looks awful, and he was screaming!" She answered.

"Maybe he was having a bad dream. You know how kids are."

"Are you listening to me? I said he looked awful. Bad dreams don't make you look awful. I touched his forehead to see if he had a fever, and his skin felt funny."

JW could hear the urgency in his wife's voice and climbed out of bed. As soon as JW saw his son, he knew something was wrong.

"What's wrong, little buddy?" he said, sitting on the edge of the bed, laying his hand on John's chest in an attempt to comfort him.

John opened his eyes, but didn't say anything. JW had seen shark eyes before when he worked as a whaler on the Hercules, sailing the Atlantic Ocean. JW looked into his son's eyes. They were two pools of endless black. Hard and flat with no reflection. Like shark eyes.

JW stilled dressed in his pajamas, wrapped his son up in his blanket, and said to his wife, "Get the keys to the truck!"

JW carried John downstairs with Mary in front of him. She stopped long enough to grab the truck keys on a peg next to the front door.

JW carefully sat John in the front seat of the truck. Mary got in next to John and closed the door. JW ran around the front of the truck and hoped in. He started the truck and gunned the engine. The four-wheel drive shot out of the driveway and down the street. JW raced to Anchorage's General Hospital driving as fast as he could, leaning on the truck horn whenever his progress was slowed by the occasional vehicle. When they reached the hospital, JW carried John inside.

"My son needs help!" he shouted to a nurse sitting behind a chest-high counter, oblivious to the other people already there waiting.

"What's the problem?" asked the nurse calmly, as she was trained to do. She knew it was important for her to stay calm to lower the stress level when interacting with the public who were already in a stressful situation for themselves and/or their loved ones.

"I don't know, but he's really in bad shape. He's my son," JW said.

The nurse stood up to get a better look at John. It was obvious from his complexion that something was wrong. She picked up the phone and said, "We may have a possible code red involving a child." After talking for a few more minutes, she hung up. Several nurses appeared, rolling a gurney. JW laid John on the gurney. The nurse behind the desk said, "One of you will have to stay here so I can get the patient's name and other information for treatment and possible admission."

"You go, honey, and stay with John. I'll take care of this other stuff and meet up with the two of you later," JW said.

Mary nodded her head in acknowledgment, squeezing her husband's hand.

She followed the gurney as the nurses wheeled John into a closed-off cubicle. "Please wait here, ma'am."

Mary sat on a bench and waited as she saw more nurses enter the cubicle. She looked along the corridor. There were other people there in various emotional states. Some were crying quietly. Some looked very sad. Some just stared straight ahead.

Inside the cubicle, the nurses moved swiftly and efficiently. John's blood pressure and temperature were taken. A blood pressure reading contains two numbers: systolic pressure and diastolic pressure. Systolic pressure is the top or first number in a person's blood pressure reading that indicates the pressure within the arteries when the heart pumps out blood. Diastolic pressure is the bottom or second number, indicating the amount of pressure being exerted on the arteries when the heart is filling with blood. A normal blood pressure reading is 120/80. John's was 90/80, which is dangerously low, signaling an oncoming stroke, shock, or complete body shut down without an immediate injection of adrenaline.

Hypothermia is a medical emergency term that occurs when the body loses heat faster than it can produce heat, causing the body's temperature to drop dangerously low. A normal body temperature reading is 98.6. John's body temperature was down by 4.6, and all of his internal organs were starting to fail.

Code blue meant a possible poisoning. A code red meant a life-threatening emergency. John's code was changed from blue to red. Two men wearing white smocks rushed into the cubicle where John was being treated. They were obvious doctors. Mary saw them and wondered what was happening.

"What's going on?" asked JW, joining Mary on the bench.

"I don't know. No one is telling me anything, and just before you sat down, I saw two doctors run into John's cubicle," answered Mary, pointing with her finger at the cubicle John was in. JW could hear the concern in his wife's voice. They waited silently with JW's arm around Mary's shoulders to comfort her.

They lost sense of time until JW saw a doctor exit the cubicle and stood up. "Excuse me. My name is John Wainwright, senior, and this is my wife, Mary. That's our son John, who's being treated there. We brought him here hours ago, and since that time, no one has told us anything. What's going on?"

"One moment, please," the doctor said, and he disappeared back inside the cubicle. JW and Mary looked at each other. After a few minutes, the doctor returned with another doctor in tow.

"I'm sorry for not introducing myself the first time you spoke. There was an emergency that needed my attention. My name is Dr. Stevens, and this is my colleague, Dr. Meyers." Everyone shook hands, and afterwards, Dr. Stevens continued, "We've been treating your son, and honestly, we're baffled about his condition. Neither of us have ever seen or read anything about the symptoms your son is displaying. We know this is a very stressful time for both of you, but we need to ask some questions. When was the first time John seemed different from his normal self?"

"About two weeks ago, he had a nagging cough," answered Mary. "We thought it was just one of those summer colds kids sometimes get or an allergy. I gave him some cough syrup for the cold and an allergy pill for the allergy if that was what it was."

"Do you know the brand name of the cough syrup you gave him, and what kind of allergy pill it was?" asked Dr. Meyers, speaking for the first time.

"It was a generic brand name," answered Mary, pausing as she thought a few seconds, and then said, "Delsym, and the allergy pill was Sudafed."

"Do you know if any of the other children John may have come into contact with had this cough as you described it?" asked Dr. Stevens.

"No, but I can check with the other mothers in the neighborhood," answered Mary.

"Please, that would be a big help. Also, I would like for you to write a list of everything your son has eaten or what he might have eaten, like favorite snacks in between meals they think parents don't know about. Have your son or any of you visited or been to any strange places out of the ordinary?"

"My husband and I are homebodies. John is an adventurous eleven-year-old kid. He probably goes to all the usual hangout places around our neighborhood where other kids his age go."

"What about the family's medical history? Any diseases or hereditary illnesses?"

"My side of the family is prone to high blood pressure," answered JW, speaking for the first time.

"There are no medical issues on my side of the family that I am aware of," answered Mary. "My parents had me late in life, and they're both still alive. These questions are scarring me. You said my son's condition was baffling, but you didn't say how or in what way."

The doctors glanced at each other. "Your son's…" Dr. Stevens said, trying to sound hopeful when there is none before continuing, "Larynx has become stiff and hard, so he's unable to tell us what is going on with him. His blood pressure and temperature are dangerously low. His heart rate is starting to slow down. His skin is turning a grayish color, as you saw, and his eyes are no longer dilating. Your son's internal organs are also becoming stiff and hard, making it very difficult and laborious for them to perform their assigned functions. In essence, your son's body is starting to shut down. We understand why. It's the cause we find baffling for this kind of complete failure."

"Is there anything that can be done to reverse whatever is happening to him?" Asked Mary.

"If we knew what was the initial cause of his symptoms was, then perhaps. All we can do now is try and stabilize him," answered Dr. Stevens.

"Can we see him now?" Mary said quietly.

"Sure," Dr. Meyers said. As he turned to leave, JW touched Dr. Steven's arm. He stopped as Dr. Meyers and Mary continued on walking into the cubicle.

"Is my son dying, Doctor? You don't have to be gentle with me. I'm an old whaler, and I've seen death many times."

"Yes, he is. I'm sorry. We'll make him as comfortable as possible."

"Thanks Doc. My wife and I appreciate that. I want to be with them now."

"Of course," Dr. Stevens said. "Please follow me."

When they entered the cubicle, Mary was sitting in a chair. She was leaning over the bed. She held her son's hand pressed against her cheek. Dr. Meyers was gone. "Why is it so dim in here?" JW asked Dr. Stevens.

"We think the lights hurt his eyes."

JW remembered how opaque his son's eyes had looked. John's body looked dwarfed by the various machines connected to it, humming in the silence. The iridescent lights glowing from the machines in the dimness gave the cubicle a surreal feeling. "If you have any more questions, please have one of the nurses page me, and I'll make myself available," Dr. Stevens said as he left the cubicle with the nurses trailing behind him. They understood it was a private moment.

During the dawn hours of the morning, John was transferred to the intensive care ward of the hospital, where he lingered on, clinging to life for three days, each day more painful than the day before, before succumbing to death. His body was so hard and stiff that his weight had almost doubled from a hundred and thirty-five pounds to two hundred pounds. The cause of death was listed as unknown.

Chapter 2

Ronald Jordan came from a long line of whalers. It was rumored among his family that his great-grandfather served as a guide for Admiral Robert E. Pearly and Matthew Henson's four-member Inuit Eskimo expedition team when they last set out to explore the Antarctic. No one really knew for sure if it was fact or fiction. It was just a rumor, like most families have, that gets passed down from one generation to the next.

When Ronald was born in 1985 at Fairbanks General Hospital in Alaska, his proud parents, Frank and Martha, affectionately shortened his name from Ronald to Ron. There was no doubt Ron was destined to be a whaler like his father and his father's father before him.

Ron's training started early. He was twelve when he first started going out with his father. At first, they would just go out fishing, standing on the banks of Lake Ural. After that, his father started taking Ron out on the lake in a small row boat to acclimate Ron to the movement of the boat as the water brushed against the hull of the small boat. Ron's father taught him the names of all the indigenous marine life that swam in the lake.

Two years later, Ron sailed with his father for the first time on Martha. It was a whaler ship named after his mother. The ship was massive; from bow to stern, it measured two hundred feet, and from starboard to port side, it measured one hundred feet. At the front of the ship was a huge hydraulic crane used to hoist whales out of the water. On either side of the crane was a harpoon gun. In the center of the ship was the wheelhouse. Four thick panes of glass windows allowed the ship's captain, Ron's father, a panoramic view of the ship and whichever of the seven oceans that they were sailing on, following the migration of whale patterns. All whales in a population may not migrate. However, when they do, whales undertake some of the longest migrations on earth, swimming many thousands of miles over months to breed in tropical waters. Whaler ships were built for long, arduous journeys. Next to the wheel house was the mast for unfurling the ship's canvas sails for when the ship's massive engines had to be silenced to calm the whales they were stalking. At the back of the ship was the ship's huge arrow-head-shaped anchor. Next to the anchor were meat containers half packed with ice, and next to the containers were stand-up metal lockers where an assortment of knives, saws, hooks, and other very frightening-looking objects were stored. On the starboard and port sides of the ship were cargo netting and small motor boats. The ship had two other decks below the main deck. The second deck was the ship's sleeping quarters, small kitchen, dining hall, showers, and bathroom. The bottom deck housed the ship's laundry room and powerful diesel engine. The ship had a fifteen-member crew. One was a cook, and one was the ship's mechanic. The remaining thirteen members of the crew were seasoned whalers. They were all tough, strong, loyal, and capable young men.

Ron hated his first time out in the ocean. He learned that being on a ship was nothing like fishing in a small row boat. As massive as Martha was, he could still feel the ship sway in the open ocean like it was a toy. The water seemed like it had no end. The wind blowing off the water made the air freezing cold and wet. They stayed out for hours without spotting any whales. What really bothered Ron the most was constantly being wet.

Perhaps it was due to genetics. On Ron's third time out, he had grown acclimated to the ocean. They were sailing the Arctic Ocean when one of the crew members, Jake, yelled out, from behind one of the harpoon guns, pointing, "I see a big one!"

Ron went to the front of the ship, and for the first time in his life, saw a live whale in all its majesty in its natural habitat. The whale was huge. Almost as huge as the ship. The whale bobbed in and out of the water effortlessly.

"Ron, come here," Jake said, moving away from behind the harpoon gun.

"Have you ever shot one of these babies before?"

"No," Ron answered.

"Okay. Listen up," Jake said. "This is not a toy. I want you to stand here. Put both hands on these levers. Point and squeeze them at the same time. When the harpoons release, you're going to feel a jolt. Keep holding the gun steady. Aim for center mass. That's the body of the whale. In the middle."

Ron stood behind the harpoon gun, just as Jake had instructed. He braced himself, took careful aim, and waited. As soon as the whale came up for air, Ron squeezed both levers of the harpoon gun. The harpoon made a whooshing sound as it sailed through the air on its deadly course. Ron felt the jolt of the gun. He held the gun steady. The harpoon disappeared into the ocean. Ron thought he had somehow missed.

Suddenly, he felt the ship jerk forward.

"You got her boy!" Jake shouted.

"What do I do now?"

"Nothing. We'll go for a ride. Pulling a ship this size will take some of the fight out of her," Jake explained.

The ship's engine was cut, and the sails were unfurled to create more drag and resistance. It still took most of the day for the ship's momentum to start slowing down. The ship's huge crane lowered its grappling hooks into the water. The ship's stern tilted forward from the weight of the whale. Once the whale was hoisted in the air and out of the water, and before its massive body was laid on the deck of the ship, Ron's father took a picture of the crew standing next to the whale. A crew member named Lester took a picture of Frank and Ron standing next to the whale to memorialize Ron's first whaling. Frank was proud of his son. The family's legacy was safe. Afterwards, the whale was laid on the deck of the ship. The crew went to work applying their deadly skills, gutting the whale, slicing and dicing, and taking out the whale's internal organs. The work was hard, dirty, and very bloody. Those parts of the whale that were not salvageable were tossed back into the ocean by the sharks trailing the ship after smelling the whale's blood. The remaining parts of the whale, the fat, would be stored in the ice containers to be sold on the market and then turned into grease, oil, and other lubricant products. The meat would be stored in the ice containers after first being salted for preservation and sold on the market. The ship's crew members would also take some of the meat to feed their families.

Growing up working alongside his father and the ship's crew members made Ron think his father was invincible. The way the crew respected him, responded to his commands, and his knowledge of the ocean all added to his belief until one very cold Saturday winter night in 2019.

Alaska is considered the last unexplored frontier where women were scarce and in high demand. Women came to Alaska from all over the world to meet that demand. The majority of the women hoped to find that special man to marry, have kids with, and settle down, while others hoped to find men who would fatten their purses before leaving looking for their next victim. Like most whalers, after a long day and sometimes months out at sea or in the ocean, crew members would hit Alaska's bars, ready for the attention of any woman who made themselves available. On that fateful night, it was speculated that Frank, after docking Martha in Fairbank's harbor, had gone to the "Drop In" for a few drinks. Someone inside the bar must have spotted Frank paying for his drinks with a wad of money. Frank was found lying face down in the snow next to his truck. Someone had stabbed him in the back as he climbed into the cab of the truck. The truck door was still open. There was one set of foot prints that led back inside the bar. Frank's killer was never caught. That was ten years ago. Ron's mother remarried and lived alone in the same house she had shared with Frank when he was alive. The joy of her life now was Ron and his two daughters. The oldest was named Brandy after the mega hit "Brandy" by the singing group Looking Glass. She was seven, and her sister Bernice was named after Ron's wife.

Ron had inherited Martha along with the ship's fifteen-member crew. They were older than when he first met them, but still tough and strong. Their knowledge about the sea, ocean, whales, other aquatic life, and the ship made them invaluable. Ron didn't have any siblings. Martha's crew was the closest he would ever know to sharing that kind of familial relationship. Ron had expanded the family business to include seal hunting to boost profits.

They had been out in the open ocean for a week when Ron started coughing. It was a dry persistent cough. Developing a cold was a hazard that went along with being out on the open waters, so Ron nursed his cold with hot coffee and "Halls" cough drops from the ship's galley to break up the phlegm he felt building in his chest.

They had three good days of hunting. The ship's meat containers were filled with seal meat, fat, and skins. It would take the ship four days of sailing north up the Pacific Ocean and the Gulf of Alaska before Martha would reach Anchorage, where the ship would dock.

On the seventh day out in the ocean, Ron knew there was something different about his cold than past ones. The cold had gotten progressively worse, and it had spread throughout the ship with the same level of progression. Two of the crew members were completely bedridden. Some moved lethargically on the upper and lower decks, while others seemed to have collapsed in the middle of whatever task they were performing. A few of the men were scattered with their backs pressed against the starboard and port side of the ship, their heads hung low, their eyes open, staring into nothingness, their faces blank, devoid of any expressions or emotions.

Ron was sitting in the captain's chair inside the wheelhouse. He couldn't remember the last time he had gotten up or left the confines of the wheelhouse to check on the crew. He felt stiff and weak, like all the energy was being slowly drained out of him. It was becoming more and more difficult to feel human.

Ron slowly, and with great effort, picked up the ship's radio mic.

"This is Martha. Mayday! Mayday! Over!" In his mind, Ron was trying to shout out the maritime emergency call for help, but he knew his voice sounded weak and frail. He could barely hear his own words.

"This is the coast guard. Where are you? What is the nature of your emergency?"

Ron heard the response and tried to answer, but he just didn't have the strength. He could sense more than feel his lips moving, but there were no words coming out of his mouth. His fingers felt like sticks of wood. The radio mic slipped from his fingers.

"Come in. What is your location, and what is the nature of your emergency?" There was no response. Pvt. Chris Owens knew something was wrong.

"Sergeant, I think we have a situation here."

Sgt. Howard Gregory was Private Owens's immediate supervisor inside the coast guard command center. "What do we have?" He asked, standing behind Private Owens, looking at the bank of blinking monitors.

"I just received a distress call from a ship identifying itself as Martha. When I radioed back for their location and nature of the emergency, there was no response."

Sergeant Gregory knew from experience that anything happening out in the water was of the essence. Like all distress calls, they were recorded, and the radio transmissions were automatically traced back to their place of origin.

Both men looked at the bank of monitors with their black faces and white iridium-sweeping hands. Private Owens pointed to one of the monitors and said, "It looks like the call originated from the Pacific Ocean near the bottom of the Gulf of Alaska."

"What's the closest ship we have near them?"

Private Owens consulted another monitor before answering. "The North Star, sir."

"Okay. I want you to radio them and give them Martha's coordinates. I'll get on the phone and get a chopper in the air."

Private Owens got on his mike and began executing Sergeant Gregory's instructions while Sergeant Gregory walked to his office and picked up the phone.

Seven hours later, Alaska's Coast Guard aerial helicopter with a five-member crew spotted Martha. The ship was sailing erratically. Lt. Rick Daly was the helicopter's pilot. Sgt. Warren Beasly was his copilot. Patrick Allen, whom everyone called Pat, and Fred Parker were the wingmen. Amy Ellison was the crane operator. They were all privates. All the crew members wore orange florescent flight pants and jackets and were equipped with radio headsets attached to their helmets, a tool belt with a flash light attached to it, and a forty-five automatic side arm.

"Hello Martha. This is Lt. Daley of the coast guard. Come in, over!" There was no response as the helicopter hovered over the ship. It was dark in the early morning hours. The light of dawn had not broken over the horizon.

Lieutenant Daley shined the helicopter's powerful search light on the ship's deck. He could see silhouettes strewn along the starboard and port side. Because of the ship's mast, Lieutenant Daley couldn't risk bringing the helicopter down lower for a closer look.

"Pat, get ready. I want you to go down and take a look."

"I'm on it, Lieutenant," Pat said. He had seen the silhouettes too. Private Allen adjusted the crane cable to his harness that he wore over his flight jacket while Private Parker opened the helicopter bay doors. The frigid air blowing off the gulf filled the helicopter interior.

"Ready!" Shouted Private Parker.

Private Allen, with his back toward Private Parker, gave him the thumbs-up sign and stepped out into the darkness.

"Keep it coming!" shouted Private Parker to Private Ellison over the helicopter rotary blades as he held the cable steady in his gloved hand while beckoning with his other hand for Private Ellison to keep lowering the cable.

When Private Allen touched down on Martha's deck, he yanked hard on the cable one time.

"Stop!" shouted Private Parker.

Private Ellison stopped the crane.

Private Allen unhooked the cable from his harness while his eyes swept over the ship's deck. He took out his flash light. The light looked eerie as it cut through the dark silence of the night. Private Allen could see men sitting along the starboard and port sides of the ship. Their heads hung low, with their chins resting on their chests. Their eyes were open. Private Allen shifted the flash light from his right hand to his left hand and took off his right glove. He felt a pulse on each man. There were none.

"Lieutenant," Private Allen said into his headset mic. "The situation is bad. We may have a ghost ship. So far, I've found what looks like six crew members. They're all dead. Over."

The radio was silent. Private Allen knew Lieutenant Daley was processing the information. After a few seconds of silence came, "Can you determine the cause of death? Over." The term ghost ship was code for dead ship.

"Negative Sir. I didn't inspect the bodies in detail, but there were no signs of blood, projectile wounds, or any external trauma to any of the bodies that I could see."

More silence, then, "Check the rest of the ship and get back with me ASAP. Over."

"Over, sir."

"Pat."

"Sir?"

"Be careful."

"Roger that. Over."

Private Allen upholstered his weapon and began searching the ship cautiously, with his gun in one hand and a flash light in the other. Private Allen wasn't anticipating any trouble, but he knew being out on the open water did strange things to a sailor's mind, and he wanted to be prepared for the unexpected.

Private Allen knew his way around a ship. He didn't waste any time, as he moved methodically, starting on the top deck. He entered the wheelhouse. Ron was still in the captain's chair. The radio mic was dangling from its cord. Ron's eyes were focused on a picture taped to the wheelhouse's front glass window. The picture depicted a whale hoisted in the air by a huge crane. On one side was a man, and on the other side was a little boy. Private Allen looked at the picture, then at Ron, wondering who the man and the little boy were. Private Allen went below deck. There were four more bodies sitting at the kitchen thick wooden table. He didn't have to check their pulses to know they were all dead. There was a battered steel cup in front of each man. Private Allen picked up one of the cups and sniffed. Whiskey. Private Allen returned the cup to the table and looked at each man. They were in various positions. One of the men had his head tilted back as if he were looking at something on the ceiling. Another man had his arms out before him with his head hung low and resting on his chest. Two others were slumped forward. Private Allen wondered if the men had gathered at the table for their last drink, knowing death was sweeping through the ship like a pestilence. Private Allen found four more men in the ship's sleeping quarters. Two of the men were lying in their bunks. The other two were kneeling at their bunks as if they were praying.

Private Allen went back to the top side.

"Private Allen to Lieutenant Daley. Over."

"I'm here. What's going on, Pat? Over."

"It's confirmed, sir. We have a ghost ship. I counted fifteen bodies."

Dawn was starting to break over the horizon. "Is the ship operational? Over."

"There doesn't appear to be any structural damage. Over."

"I'll take that as a yes. The North Star is on a course to intercept you. I want you to slow the ship down until she arrives. The North Star's crew will board the ship and help you bring her into port at Anchorage. Stay with the ship until you get further instructions. Over."

"Roger that."

"Over and out."

After Lieutenant Daley signed off, Private Allen returned to the wheelhouse. He slowly and respectfully removed Ron's body from the captain's chair. Private Allen reduced the ship's speed and then began rummaging through the scattered papers on the ship console until he found what he was looking for. It was the ship's log book. He opened the book and started reading.

It was early in the afternoon when Private Allen finished reading the log book. He knew the ship was named Martha and that she had a fifteen-member crew. He knew the men's names, but not which name belonged to which man except the captain. Private Allen knew the man he removed from the captain's chair was Ron and that he was the ship's owner. The log detailed how the ship had sailed the Arctic Ocean, catching seals. A member of the crew named Jake was the first to start coughing, catching seals in the Arctic Ocean. A member of the crew named Jake was the first to start coughing. Everyone thought the coughing was just another typical cold at first, until after a few days, no one was getting any better. Jake succumbed to death first, and then one by one, the men started dying.

Private Allen finished reading the log book and began looking through the wheel house cabinets until he found what he was looking for. A first-aid kit. Every ship was required to have one by maritime law. Inside the kit were three flares, a flare gun, bandages, a bottle of iodine, two bottles of water, and two chocolate nutrient bars. Private Allen had forgot the last time he had eaten. The nutrient bars tasted bland, but they were packed with protein, and along with the water, he felt replenished. After satisfying his hunger, Private Allen raised the North Star on the ship's radio.

"This is Private Allen of the coast guard aboard the Martha. What is your ETA? Over."

"This is Captain Travis of the North Star. We are approximately one day away from intercepting you. Our present location is south of the Bering Sea. Over."

"Have you been informed about the status of the ship's condition? Over."

"Affirmative. Has anything changed? Over."

"Negative. Once we get to port, I'll need all the men you can spare to dock the ship. Over."

"We'll be ready to offer you whatever assistance you need. Over."

"Roger that. Over and out."

Private Allen returned the mic to its cradle. There were no icebergs or other ships near Martha. Private Allen occupied his time by taking a cat nap during the day. Nighttime was the most dangerous time for ships because of the reduced visibility, so he stayed awake at night reading the ship's maps and tracing Martha's movements.

The North Star intercepted Martha at noon the next day. Six men departed the North Star and rowed up to Martha's port side. They climbed up the cargo netting and boarded the ship. Private Allen met them, and after introductions were made, they took up stations and began sailing the ship following the North Star.

It took four days and three nights to reach Anchorage's port. The ship was docked. Lieutenant Daley, Sergeant Beasley, Privates Parker, Ellison, and other members of the coast guard were waiting. The bodies of the dead men were taken off the ship and loaded into the back of a waiting morgue truck. The bodies of the dead men, Private Allen, and the six men from the North Star were all taken to the city's general hospital for examination. Private Allen and the six men from the North Star were examined first. They were all given a clean bill of health except for what the doctors described as a nagging cough. They were told to drink plenty of hot liquids, given some generic cough drops, and released.

It took a month to complete the autopsies on Martha's fifteen-member crew and another month to complete the toxicology. The toxicology reports were all negative for any poisons. The autopsy reports were baffling. They revealed the men's internal organs had all shut down prior to death and were almost stone-like in their texture and rigidity. The cause of death for each man was listed as unknown.

Chapter 3

The Center for Disease Control and Prevention in the United States is the biggest and most technologically advanced health organization in the world. Most people only know the organization by its acronym, CDC. The CDC is located in Atlanta, Georgia, on 1600 Clifton Road on a sprawling 65-acre plot of land. Inside the massive multilevel complex were the world's most dangerous, infectious, and deadliest diseases and antidotes. There were research laboratories where scientists worked on cures for known diseases such as cancer, E. coli, Ebola, and many others.

The first Geneva convention held in Geneva, Switzerland, covered the sick and wounded in war and concluded in 1864 at a conference by the Swiss government at the urging of the International Committee of the Red Cross.

The convention was amended and expanded in 1906. In 1929, two more conventions covering the wounded and prisoners of war were signed. Outraged at the treatment of prisoners and civilians during World War II, a third convention concluded on August 12, 1949, with a ban on the use of chemical and biological agents during times of war. Nonetheless, there were still clandestine research laboratories conducted by military scientists who worked on highly classified projects that no one except those with the highest level of security clearance had a privilege to. That still didn't stop the rumor mill from churning out rumors and conspiracy theories. Some suspected the scientists were trying to develop a disease that would annihilate the Black race. Others suspected the scientists were trying to develop a chemical or biological weapon that would give the United States armed forces a tactical advantage on the battle field in spite of the Geneva Convention accords. The CDC is a nonpolitical health organization. They never confirmed or denied any of the rumors beyond their mission statement of "No Comment."

In every hospital in the United States, there is a CDC liaison person whose job it is to document and report back to the CDC any causes of death or health abnormalities for further examination and evaluation.

Teresa Lemons was the CDC liaison person at Anchorage General Hospital. She was very efficient at her job.

When the medical reports, autopsies, toxicology reports, and death certificates of Martha's men reached her desk, Teresa immediately recognized the abnormalities. She cross-referenced all of the reports and death certificates with John Wainwright. Teresa remembered the case so well because of the way he had died, and he was so young. She sent all of the information, along with her report, to the CDC.

The summer of 2027 will be Antarctica's warmest, beating the record set in 2017.

In 1957, the United States established the Weather Research Center, or WRC, as it was called, at the top of Vinson Massif, which is the highest point in the sentinel range near the base of the Antarctic Peninsula. The research center location was specifically chosen because of the two large ice shelves, the Filehner Ronne ice shelf and the Ross ice shelf, fed by glaciers, snow accumulations, and strong winds called the roaring forties and furious fifties that blew from west to east.

That summer, the winds blew from east to west, taking along with them vapors of water from melting glaciers.

During the International Geophysical Year of 1957–1958, twelve nations established ostensibly called weather research centers on the Antarctica that were in reality intelligent gathering satellites designed to monitor the other naval movements in the Southern Ocean. For purposes of appearance, the research centers did perform their official functions. When the winds shifted that summer, the research centers noted it along with the warmth in temperature.

When the research workers, who were all scientists in various fields of discipline, started getting sick, everyone attributed the coughs, scratchy throats, and sneezing to the change in temperature.

They thought it was just a cold that would pass. They fought it with the traditional remedies of hot liquids, cough syrup, and bed rest. After two weeks of persistent coughing, the scientist knew something was wrong. They were men and women trained to observe.

When a status report of what was happening in the Antarctic reached the CDC, a medical research scientist was dispatched to gleam further information.

A CDC helicopter dropped Dr. Sheila Davalloo into a nightmare. As Dr. Davalloo ran from the helicopter, she wondered why no one was there to greet her. She felt a little disrespected. She made a mental note to speak with the WRC director about her reception. Dr. Davalloo reached the WRC door and opened it. She froze. Dr. Davalloo looked around inside as far as her eyes would allow her to pull focus. There was no doubt that everyone inside the hub was dead. She had been around dead bodies often enough to know the stench of death.

"What's the holdup?" asked one of the helicopter crew members as he attempted to push past Dr. Davalloo.

"Stop!" She said, holding her arm out to block the entrance inside the WRC.

The two helicopter crew members accompanying her, Arnold Laney and Pete Johnson, stopped as they looked over Dr. Davaloo's shoulder.

"Are they all dead?" asked Arnold.

"I'm afraid so," Dr. Davaloo said matter-of-factly. "The real question is, what caused their death? I don't see any weapons from here."

"What are we going to do?" asked Pete. "We can't just stand here and not go in. Someone may still be alive and need our help."

"I agree," Dr. Davalloo said. "We have to return to the helicopter first."

Once they were back inside the helicopter, Pete explained to the pilot, Todd Baskins, and copilot Alex Payton, what they had seen inside the WRC while Dr. Davalloo busied herself in the back. "I'm ready," She said.

They all turned around. Dr. Davalloo had changed into a hazmat suit, complete with a respirator. "Is there an extra walkie-talkie on board?" She asked.

"Here," answered Todd, reaching underneath his seat and giving Dr. Davalloo a walkie-talkie.

"The frequency is already set. You may want to check the batteries, though. I don't know how long I've been sitting on that thing."

"Testing, testing," Dr. Davalloo said. Her voice was transmitted clearly on the walkie-talkie.

Pete, Arnold, and Dr. Davalloo stepped out of the helicopter and walked back to the WRC. The first time they made the walk to the WRC, they didn't notice the crunching sound of the snow and ice beneath their high-top snow boots or how blue and clear the sky was.

"Wait here. I'll stay in radio contact," Dr. Davalloo said, once they reached the door of the WRC.

"If you need us, don't hesitate," Arnold said.

"Thank you," Dr. Davalloo said. She was the only one equipped with a hazmat suit and respirator.

The WRC was shaped like an igloo, with three main corridors that extended from the hub. In the middle of the hub floor were the center operation stations. There was a round table that was set like an island in the middle of the hub floor. Dr. Davalloo speculated that most of the work and discussions were done at the table. On the table were stacks of paper, all sorts of writing utensils, compasses, maps, and other instruments she wasn't familiar with. There was a shortwave radio console on one side of the igloo. There were two men sitting at the console, slumped over. One of the men had his hand on the mic as if he were ready to speak. Dr. Davalloo looked around at the men's faces. They were all grayish.

"I'm in what appears to be the command post. So far, I've found three dead bodies," Dr. Davalloo said into her walkie-talkie.

"Can you determine the cause of death or what happened in there?" asked Pete.

"Negative to both questions. All I can say for sure is that the bodies I've seen so far, their deaths don't appear to be the result of violence. Something did happen in here. I don't know what. I'm going to look around some more."

Dr. Davalloo clicked off. She walked along the igloo's right-side corridor. At the end of the corridor was a sliding door that opened up into the WRC sleeping quarters, bathroom, and showers. There were nine sleeping cots across from each other. All the cots were empty and disheveled, except three. They had bodies in them. They all had knit skull caps. Their thick, heavy wool blankets were pulled up to their chins. Dr. Davalloo pulled back each blanket. The men were all fully dressed in their winter clothing, as if they were extremely cold. She looked around, and just like the other bodies, there were no signs of violence.

Dr. Davalloo continued her investigation by walking along the center corridor. At the end of the corridor, there was another sliding door that opened up to the WRC sick bay. Dr. Davalloo found six more bodies. Three of the bodies were men. They were all fully dressed. Only one of them had his blanket pulled up to his chin. The other three bodies were women. Dr. Davalloo could tell by the insignia on their clothing that two were nurses and one was a doctor. Death had claimed the nurses while they sat in their chairs at the bedside of two of the men. They were holding their hands. Dr. Davalloo wondered if the nurses were simply comforting them and themselves or something deeper. She continued to look around. Dr. Davalloo found the doctor sitting at her desk inside the bay office. There were papers scattered over the desk. She made a mental note to read the papers later.

Dr. Davallloo walked along the last corridor on the left side of the igloo and passed another sliding door that opened up to the kitchen. Dr. Davalloo looked at the food in the refrigerator. The food was still edible, which meant death wasn't long in coming. There were three more dead bodies. They were all men and fully dressed as the others. They were sitting at the kitchen table. In front of them were styrofoam cups. Dr. Davalloo picked up a cup and looked inside. Black coffee that was now cold. At the back of the kitchen was a door with a window. The window was covered with frost. Stalactites hung from the window frame. Dr. Davalloo was just barely able to make out the silhouette of a shed.

"Pete. Arnold," Dr. Davalloo said into her walkie-talkie.

"Go ahead, doc. We're here," Pete said.

Dr. Davalloo bristled a little at the word doc. It was too familiar and unprofessional. In time, she would have to speak with Pete about that. For now, she lets it go. "I'm in the kitchen. I found three more dead bodies in the sleeping quarters and six more in the sick bay. Three of those were medical personnel. They all seemed to have suffered the same fate. In the back of the igloo, there's some kind of shed. I need someone to go and check it out."

"We're on it," Pete said.

"When you open the door, don't go in for a few minutes. Let it air out first before going inside. If you see any bodies in there, don't touch anything and get out immediately. I'll be at the front of the WRC waiting."

"We'll see you there," Pete said, and he clicked off.

They walked around the back of the WRC and saw the shed. It was made out of corrugated tin with a slanted corrugated roof for excess snow and rain runoff. The door was stuck with snow and ice. It took both men pulling together with all the strength they could muster to open the door. They waited a few minutes, letting the shed air out, before entering inside.

The shed had a single bare light bulb that hung down from the ceiling that was still on, casting the interior of the shed in an eerie, shadowy glow. There were three industrial-sized generators. Behind the generators were pallets stacked with drums filled with diesel fuel. There were picks, shovels, and tools laying around, as well as tool boxes. There were bags of salt and two snowmobiles. There were no bodies.

When they returned to the front of the WRC, Dr. Davalloo was waiting for them. Pete explained what they had found. When he finished, there was an ensuing silence.

"What do you think?" asked Arnold, breaking the silence.

"I don't know what to think. I'm going to stay here and try to figure out what happened. I wrote down a list of things I'm going to need. I want the both of you to fly back to the CDC, pick up the things on this list, explain what's going on here, and get back here ASAP."

"You think that's smart? I mean, staying here by yourself?" Pete said, with concern in his voice.

Dr. Davalloo looked at Pete. He can be brusque with the "doc" comments, she thought, but he can also be thoughtful. "I'll be alright. Everyone here is dead. It's what killed them that concerns me. There's a short-wave radio inside that still works. I can contact the helicopter and CDC if something should happen."

"We better get started, Doctor. The sooner we leave, the faster we can get back," Arnold said. Dr. Davalloo watched as the two men walked quickly to the helicopter.

Dr. Davalloo watched as the helicopter lifted off and flew over the horizon before approaching the WRC. She opened the door and let the WRC air out for a few long minutes before going inside. When she stepped inside, she made a silent prayer and took off the suit's head piece that was attached to the respirator. Dr. Davalloo waited, breathing smoothly and evenly. After a few minutes, she took several deep breaths. Dr. Davalloo knew what she was doing was risky. It was a calculated risk. A thesis was starting to form about what happened that killed the people inside the WRC.

Dr. Davalloo looked around inside the hub, trying to decide what she should do first. There was nothing she could do for the dead. They would have to wait until their bodies were returned to the CDC for a thorough examination. Dr. Davalloo decided her time would be best utilized reading the daily logs of the WRC operational activities. The logs were placed neatly on the table among the disarray of other papers, as if someone had taken special care to place them there.

Dr. Davalloo sat down at the table and began reading. It was dry, tedious reading that dealt mostly with detailed weather conditions. Wind currents, wind speeds, and temperature shifts. There were other terms she wasn't familiar with. Squall lines, bora, and occluded fronts. Dr. Davalloo made a mental note to have a meteorologist look at the logs and explain to her what the terms meant.

Dr. Davalloo didn't know how long she had been reading when she suddenly stopped. She sat still, listening. Nothing. There was complete silence. Dr. Davalloo knew from reading the logs that there were other WRCs nearby. Particularly the Russians. Dr. Davalloo knew that in the stillness of night, sound travels over great distances. Dr. Davalloo read back through the logs until she found what she was looking for. It was a name.

Dr. Davalloo walked over to the short-wave radio. She looked at the radio and adjusted the frequency before speaking into the mic.

"Hello?" There was no response. Only static. "Hello? Mr. Nikolai Kulikov."

After a few more moments of static, someone answered with a thick Russian accent.

"Identify yourself."

"This is Dr. Sheila Davalloo. I'm a pathologist. I work for the CDC."

"CDC," the voice repeated.

"Yes. It's an American health organization."

"I know about your CDC. What is your business, please?"

Dr. Davalloo paused before continuing. She didn't know how much information she should share. Dr. Davalloo decided that since the CDC was a nonpolitical organization, she would continue with that tradition and be forthright. "As a matter of courtesy and concern, I'm informing all the research facilities in the area that Mr. Bixler and all his team members are dead." Dr. Davalloo had learned from the logs that Bill Bixler was the WRC Director and Chief Scientist.

There was a long pause on the other end of the short wave radio, and then, "And what was the manner of their death?"

"That's what I'm here to investigate."

"I see. My name is Sergi Avilov. I'm with the BDC. Unfortunately, we share similar circumstances. Comrade Nikolai Kulikov and the other comrades are all dead. I'm here to find out why and how. Perhaps we can be of assistance to each other."

The BDC is an acronym for the Bureau of Disease Control. It's Russia's equivalent to America's CDC.

"I have to get approval from my superiors first."

"Of course," Sergi said. He knew all about getting approvals from superiors.

"I'll contact you again. Soon."

"Of course," Sergi said. Dr. Davalloo clicked off.

Chapter 4

Contrary to popular belief, the Civil War was not fought over the morality or legitimacy of slavery. Also, contrary to popular belief, President Lincoln never believed Black people of African descent were equal to White people of European descent on any level. The Emancipation Proclamation Act signed by President Lincoln on September 22, 1863, for which he is so heralded, only freed those slaves held in bondage in the Southern States that were in rebellion but did absolutely nothing for those slaves held in bondage in other parts of the United States that were not in rebellion.

The Civil War was a conflict between two classes of white people. The beef was over which economic structure would fuel the United States economic growth. The Southern States white-class economic system was premised on slavery. The Northern States white-class economic system was premised on industrialization with the invention largely of the steam pump. Black people of African descent were simply caught in the middle of a family feud and used as pawns. They were often used as spies by the north for reconnaissance missions in the south by disclosing valuable information on troop locations and other matters related to logistics.

In 1865, the Civil War ended with a series of surrenders. The first surrender of great significance was the surrender of confederate soldiers under Robert E. Lee to General Ulysses Grant on April 9 at Appomattox Courthouse in Richmond, Virginia. Later that same year, of less significance, General J. E. Johnston surrendered to General William T. Sherman at Durham Station in North Carolina. Ninety days later, the last of the confederacy surrendered.

Once the Southern States were defeated by the Northern States, the question arose: What to do with black slaves who were now free? The question was answered with the establishment of the Army Bureau of Refugees, Freedmen, and Abandoned Lands, which later became more commonly known as the Freedmen Bureau. The Bureau's mission was to integrate former slaves into America's society by promising them forty acres, a mule, and other far-reaching programs like social welfare in the form of rations, schools, courts, and medical care.

President Lincoln was shot to death in 1865 by John Booth, which allowed Vice President Andrew Johnson to ascend to the office of president. That same year, President Johnson appointed Civil War hero Oliver Otis Howard as Commissioner of the Freedmen Bureau. Under Howard's tutelage, the Freedmen Bureau started carrying out its mission. However, President Johnson often times clashed with his appointee over his efforts and tried to return political power back to southern whites by reneging on the forty acres of land and a mule promise.

Howard had the support of radical Republicans in Congress, and in 1867, the radical Republicans gained power in the House and Senate. When President Johnson removed Secretary of War Edwin M. Stanton from office without notifying the Senate, he went too far. In February of 1868, the house and senate impeached him on the charge of violating the Tenure Office Act, but in reality, the house was responding to his harsh opposition to congressional reconstruction. President Johnson was acquitted by a one-vote margin in the Senate. President Johnson got the point. Blacks were given the right to vote in the South, setting up elections where, for the very first time in history since Black Africans were brought to the shores of America, they had assumed political power. This period in history became known as the reconstruction period.

On November 20, 1866, ten members of various socially concerned groups, including Oliver Howard, met in Washington, D.C., to discuss plans for a theological seminary to train Black ministers. Interest was sufficient. However, to expand on that concept, it was needed to create an educational institution for areas other than

Black ministers land and an institution for training. The name of the institution was originally called Howard Normal and Theological Institute for the Education of Preachers and Teachers. On January 8, 1867, the board of trustees voted to change the name to Howard University. Oliver Howard served as president from 1869 to 1874. In modern times, Howard University is simply known as HU and is a privately chartered historically black university accredited by the Middle States Commission Higher Education with a $688.6 million endowment and "Veritas et Utilitas" as its motto, which means in English, "Truth and Services."

When Elizabeth Fehlinger succumbed to her death from a mysterious illness, her husband, Sterling Fehlinger, took the couple's only daughter, Liz, as far away from Alaska as he could for both their peace of mind and a new beginning. In Alaska, everything reminded them of Elizabeth. Sterling became a tenured professor at Howard University.

James Scott, whom everyone called Scotty, sat in the back of the auditorium listening to Professor Fehlinger give a lecture on cell biology.

"Do cells share characteristics?"

Professor Fehlinger looked at Scotty. At first, he didn't recognize him. It had been almost twenty years since Scotty graduated from HU at the top of his class and was editor of the "Hill Top," the university's newspaper. When recognition finally set in, Professor Fehlinger smiled and answered his own question. "Yes, they do. All cells share four characteristics." Professor Fehlinger held up one finger before continuing, "First, every cell is enclosed in a thin membrane which provides shape and acts as a barrier between the cell and its environment." "Second," Professor Fehlinger said, holding up two fingers, "All cells are filled with cytoplasm." "Third," up came three fingers, "Within the cytoplasm of all cells are organelles called ribosomes, which manufacture proteins for the cell, and the fourth," four fingers were up now, "All cells contain chromosomes composed of genetic material which encodes instructions for making the hundreds or even thousands of different proteins found in a cell. Any questions?"

After looking around, there were no questions. Professor Fehlinger said, "Okay, the next study will be about homeostasis and enzymes." Students quickly scribbled in their notepads as the bell sounded, ending the class period. When they were alone in the interlude before the start of the next class, Scotty walked up front, shook Professor Fehlinger's hand, and said, "It's been a long time, Professor."

"Too long," Professor Fehlinger said. "I almost didn't recognize you. Last I heard, you were appointed director at the CDC. What brings you here to annoy me?" He finished with a smile on his face. Scotty was his best student and had graduated summa cum laude.

"I wish I could say I was here because I was feeling nostalgic, but that's not the case. I need help."

"Let's sit down," Professor Fehlinger said, motioning with his hand to the auditorium's front row seats. After sitting down and turning slightly to face each other, Professor Fehlinger said, "With all the resources at your disposal at the CDC, why do you need my help?"

"Because you're still the best pathologist I know, and for what I'm facing, I'm going to need all the expertise you've acquired over the years."

"What's this about?"

"Have you been reading the newspapers? Watching the news?"

"Of course I have. The categories of human misery and suffering are endless."

"Sad, but oh, so true. However, I was speaking specifically as opposed to generally. This is different. There's something going on in Antarctica that is puzzling me."

"You're talking about the Stone Man virus?"

"Yes."

"How bad is it?"

"It's a lot worse than the public is aware of. I don't have time to get into the details here, but I've prepared a summary with medical reports and pictures. Names, dates, everything you need to know to determine what if anything you can bring to the table to help me is in there," Scotty said, nodding his head at a brief case at his feet. "You can take it home and read the information. If it's okay with you, I can come by your place and pick it up. There's some very sensitive stuff in there." Professor Fehlinger didn't respond at first. Scotty waited. He knew the professor was thoughtful, which made him the best in his field. After several seconds had ticked by, Professor Fehlinger said, "Sure. We can talk about details over dinner. If that's okay with you?"

"That's an uneven quid pro quo, and I accept," Scotty said, smiling.

"You still remember the address?"

"Yes, I do if you still live in that brown stone on 8th Street next to that candy store."

"That's the place," Professor Fehlinger said, picking up the brief case just as the first of his students started entering the auditorium. "I'll be expecting you at seven thirty."

"Until then," Scotty said, standing up. The students glanced at him curiously as he passed by them on his way out."

After Scotty left the auditorium, Professor Fehlinger started his lecture on cell biology. When he finished, Professor Fehlinger gave out the same homeostasis and enzymes course for home work as he gave to his other class.

Professor Fehlinger lived in a three-bedroom, three-level brown stone house with Liz, only a few blocks from the university. Professor Fehlinger walked the few blocks. It was his way of decompressing and exercising at the same time. Professor Fehlinger also enjoyed the smell, sights, sounds, and people in the neighborhood.

When Professor Fehlinger arrived at the brown stone, he opened the door and went inside. It was still early in the evening. Liz had followed in her mother's footsteps and became an anthropologist. She was a teacher at Cadoza High School. Liz was reclined on a couch with her feet up, going through some papers. On the coffee table in front of the couch was a cup of steaming hot lemon tea. Professor Fehlinger knew his daughter's habit was to relax by grading her students' assignments while sipping tea before starting dinner. "Honey, we have a guest. Please set an extra plate."

"Anyone I know?" asked Liz.

"You know him, but I don't know if you remember him. James Scott?"

"The name sounds familiar. Where would I know him from?"

"When I first started teaching at Howard, he interviewed me for the school's newspaper. He was a student then, and you were about twelve or maybe thirteen," Professor Fehlinger said, sitting down across from Liz on the living room recliner.

Liz put the papers down on her stomach and thought for a few seconds before answering, "Yeah, I do remember him. He was a Black guy. I was scared of him."

Professor Fehlinger knew Liz wasn't a racist. He didn't raise her that way. At least not consciously. Professor Fehlinger understood his daughter's observation. Alaska had the smallest percentage of Blacks of any state in the United States. Seeing a black person for the first time in her life was a frightening experience for a twelve- to thirteen-year-old girl. "He's the director now at the CDC. He stopped by to see me at the university. He wants me to take a look at something and discuss it over dinner. He's expected at seven thirty."

"I should get started putting something together then," Liz said, sitting up and arranging the papers she was grading into a neat stack on the coffee table.

"Nothing fancy," Professor Fehlinger said, rising up from the recliner.

"Dad, I got this," Liz said as Professor Fehlinger walked into the den.

The den was small but comfortably furnished with a desk that had a laptop computer and chair. On one side of the den was a book shelf filled with books. In front of the book shelf was a small, two-seat couch. On the other side of the den were windows covered by a drape. The center piece of the den was a cushy leather recliner that was set in front of a 25-inch color television. Next to the television was a floor-model receiver radio and stereo set. Professor Fehlinger flicked on the den light, turned the radio on to a jazz station with the volume turned low, and sat in the recliner. Professor Fehlinger reclined back, kicked off his shoes, opened the briefcase, and began reading with his feet resting on the recliner foot piece.

What Professor Fehlinger read he found very interesting, fascinating, and sad all at the same time. After reading the medical reports, Scotty's summary, and looking at the pictures of dead bodies claimed by the Stone Man virus, he sat listening absent-mindedly to the soft jazz music. After a few minutes of sober thinking, Professor Fehlinger got up from the recliner and clicked off the radio.

Professor Fehlinger sat down at the desk and opened one of the drawers. Professor Fehlinger found a bottle of Old Taylor he had stashed away and poured himself a stiff drink before turning on the laptop. It had been years since he had thought about his wife. He had busied himself with raising Liz and making sure his work in the field of pathology was par excellence.

When Professor Fehlinger began reviewing his wife's work on the laptop, the feeling of emptiness he had tried to escape came rushing back. Time did nothing to ease the pain, and there were no feelings of closure that people often talked about. The bottle of Old Taylor numbed the pain a little as Professor Fehlinger plowed on reading about his wife's hypothesis that there were primitive people who once lived in Antarctica. She had never gotten the chance to prove her thesis after her discovery of the Ice Man. Professor Fehlinger hated the then-president for cutting the Environmental Protection Agency funding, believing the ridiculous observation that there was no such thing as global warming.

Elizabeth had died shortly after that from a mysterious virus. Professor Fehlinger still remembered how she looked at death. She looked the same way as the pictures he had just seen. Professor Fehlinger was sure there was a connection, but what? What was the connection? Professor Fehlinger's thoughts were interrupted by a knock at the front door. Professor Fehlinger could hear mumbled voices, and then, "Dad! Mr. Scott is here!"

Professor Fehlinger looked at his watch. Seven thirty. "I'm on my way," he said, clicking off the laptop and lights in the den.

When Professor Fehlinger came out of the den, Scotty was shaking Liz's hand and saying, "Please, call me Scotty. You sure look a lot different than the last time I saw you."

"As you do," Liz said, blushing.

When Scotty saw Professor Fehlinger, he said, "I was just telling your daughter I remember seeing her when she was a little girl and how different she looks now."

Professor Fehlinger smiled and said, "Thank God she has her mother's looks. I hope you brought your appetite with you."

"I did," Scotty said as he followed them into the kitchen.

Dinner was spaghetti with beef meat balls, spicy tomato sauce, garlic bread, and a tall glass of water. As Liz was preparing their plates, Scotty said, "Professor, I hope you don't mind if I skip the pleasantries. I want to get down to business. If you read the information inside the briefcase, you know how urgent the situation is."

"I was just finishing up when I heard you knock at the door. What exactly do you need from me?"

"We have tissue samples taken from people affected by what the media is calling the Stone Man virus at the CDC. I want you to take a look at them and tell me what you think."

The two men grew quiet as Liz set their plates, laden with food, on the table. When she was finished, Liz set a plate for herself, and after saying grace, Professor Fehlinger said while twirling spaghetti around his fork, "Okay. I can do that. When do you want me to leave?"

"Going somewhere, Daddy?"

Professor Fehlinger looked at Scotty. He understood the look and said, "Liz, I need your father's expertise on a very important matter. Have you heard of the Stone Man virus?"

"I don't know anyone who hasn't," Liz answered, eating her spaghetti delicately. She was waiting. She knew there was more.

"What you don't know is that it's a lot worse than anyone knows, besides a very small, select group of people. Your father, and now you." Scotty told Liz what was inside the briefcase her father had brought to their house.

"Maybe I can help. I am a trained anthropologist."

It was Scotty's turn to look at Professor Fehlinger. Liz's response had surprised him. He wasn't expecting it.

"Fresh eyes could make a difference," Professor Fehlinger said thoughtfully.

Scotty knew he was looking at a gift horse in the mouth. "I can't disagree with you on that," he said. "How soon can both of you leave?"

"When are you prepared to receive us?" asked Professor Fehlinger.

"Yesterday," answered Scotty."

"I'll need at least one day to make arrangements at HU to have someone cover my classes and to pack a few things," Professor Fehlinger said.

"I'll need the same accommodations," Liz said.

"One of the perks of being the director at the CDC is that I have access to a fleet of government planes to fly me anywhere in the world I need to be. As soon as everyone is ready, let me know, and we can prepare for takeoff."

The rest of the evening was spent with small talk, with Scotty doing most of the talking about the CDC. The Stone Man virus hung like an ominous yoke around the dinner table.

Chapter 5

Pvt. Patrick Allen has loved the coast guard ever since he joined the United States Navy fresh out of high school. The Navy had afforded him the opportunity to travel all over the world, so when his two-year tour was up, he discharged with honors, married Rita, his high school sweetheart, and joined the coast guard.

When Patrick boarded Martha, he had been married for two years and was expecting their first child. Patrick felt good and excited about his life, so when he started coughing a few days later after being on Martha, he shrugged it off. After docketing Martha at Anchorage, Patrick passed a physical examination and was cleared to return to work. He was given cough drops for his nagging cough with instructions to drink hot liquids along with the six other crew members who had boarded Martha.

When Patrick returned home, Rita was happy to see him. Martha had been dubbed the Ghost Ship, which was big news in the State of Alaska. Rita had been worried sick with concern. Patrick assured her that other than having a slight cold, he was okay. Patrick returned to work the next day.

The North Star sailed back out to sea and then into the ocean on routine patrols. Patrick, along with the other six crew members who had boarded Martha, gradually grew sicker until, after a month of being out in the open waters, they were bedridden, and all the men aboard the North Star were having coughing spells. The captain made a decision to turn the ship around and return to the coast guard station.

When the North Star arrived back at the coast guard station, the crew were so sick that the men had to be taken off the ship in gurneys and whisked away to Anchorage General Hospital. Three weeks later, the entire coast guard station had to be quarantined. Everyone was sick and scared. The station had never been quarantined in its history.

Anchorage General Hospital was overwhelmed and chaotic. There were patients everywhere. Every room and cubicle were filled. Patients were lined up in the corridor against the wall in gurneys. It was becoming increasingly difficult to distinguish the live patients from the dead bodies. The noise level was clamorous as patients screamed for help. They screamed for the pain they were enduring. Children screamed in pain. They screamed and cried out for their parents. The doctors and nurses were starting to get sick.

An emergency meeting was called in the administrator's office. Present at the meeting were Douglas White, the administrator; Dr. George Awkard, head physician; and Diane Marshal, nurse manager.

"I've contacted all the surrounding hospitals, and they're all filled to capacity. They're all experiencing the same crisis we're dealing with," Douglas said. The office was soberly quiet. Dr. Awkard said what everyone was thinking. "I think we're in the early stages of a pandemic."

"No one said anything about a pandemic," Douglas protested.

"No, but I did," Dr. Awkard said, not backing down. "I know you don't want to hear it or even acknowledge the possibility. None of us do, but the facts are the facts. The real question is, if it is in the realm of possibility, and I'm suggesting that it is, what are we going to do about it?"

"I'm open to suggestions," Douglas said. He had to acknowledge that Dr. Awkard could be right. Something is making people sick.

"Before we go on record, we should wait a little longer and do some more testing for confirmation. That'll give us some time to contact the CDC and come up with a strategy to combat this thing. In the meantime, we'll say it's a new strain of the rhinovirus and start passing out face masks," Dr. Awkard said, taking over the meeting.

"If your prognosis is right and we're in the early stages of a pandemic, you're suggesting we allow possibly hundreds of people, if not thousands, to die and say it's because of a new strain of the flu!" Diane said incredulously.

"If you have a better idea, I'd like to hear it," Dr. Awkard said calmly.

"How about the truth?"

"Okay, let's see," Dr. Awkard said. "You want to announce to an already sick and scared population that they may have now contracted an incurable disease. What do you think will happen? Mass panic. At least my way will buy us some time to contact the CDC for guidance. With the resources at their disposal, they may be able to provide some insight on how to slow things down until we're able to announce we're in the process of fast-tracking a vaccine for this thing. That will give the people hope, and help is on the way."

"Dr. Awkard is right," Douglas said. "Unfortunately, people are going to die no matter what we do. People are already dying. We need the time Dr. Awkard suggests offers not only to contact the CDC, but to confirm we have an outbreak. We'll stick with the cover story it's a new strain of the flu for as long as we can. I'll contact the CDC and make them aware of what's going on here."

When Scotty had got the call the CDC was already inundated with calls coming in from those parts of the United States near coastal lines, Canada, and Russia.

Russia's state run media Prezaraka was using the outbreak as a political rallying point to accuse the United States of biological warfare while using diplomatic back channels to work with the CDC. Russia had closed off its northern borders, and its northern towns were under siege by the military. The bodies of people who had died were burned and buried in mass graves.

The Canadian government was a federal parliamentary democracy closely aligned with the United States Constitutional Federal Republic and was not as brutal as Russia with its people or information. The Canadians working alongside the CDC blamed the outbreak sweeping across its northern provinces on a new strain of the rhinovirus. The government imposed a curfew on its citizens to remain quarantined in their homes. The military was out too. Anyone found not in compliance with the curfew was arrested and taken to a tent city surrounded by barbed wire. Dead bodies were burned in their respective provinces and buried in mass graves.

Scotty felt desperate and hopeless. The CDC research lab, despite working around the clock with a team of the best scientists from various fields of discipline, was no closer to a vaccine than when the first confirmed case of the Stone Man virus was reported. After the call had come from Alaska, Scotty gathered up medical reports, toxicology reports, pictures, his summary, and notes, and stuffed them inside a briefcase and flew to Washington, D.C., to ask for help from his mentor.

Scotty watched Professor Fehlinger in the CDC research lab from behind a thick glass window with Liz. Scotty knew he had made the right decision by reaching out to the professor.

Professor Fehlinger was completely covered in a hazmat suit equipped with a breathing respirator. Professor Fehlinger was inside a hermetically sealed room, looking at tissue samples taken from people who had contracted the virus. The work was slow. Professor Fehlinger's arms were inside sleeves. Pincers allowed Professor Fehlinger to manipulate the tissue samples under a microscope. The tissue samples Professor Fehlinger was looking at were collected by a scientific team led by a renowned pathologist named Sheila Davalloo, whose work he was familiar with. Dr. Davalloo would understand the principles of collecting samples for examination that would eliminate the chances of foreign matter and cross-contamination.

Professor Fehlinger carefully, slowly, and methodically studied the tissue samples. The general mass of people not connected with the health care industry usually associate an organ as an internal body part, which it is, forgetting that the skin is also an organ. It's the body's external organ and the biggest organ. Professor Fehlinger started his examination there.

The epidermis is the human body's outer layer of skin, which gives human beings their skin tone because of a biological pigment called melanin. This layer, the first layer, doesn't have any blood vessels, which is why when there is a paper cut and it's not too deep, it doesn't always bleed. The skin cells are all dead but tightly attached. The second layer of skin is called the dermis, where sweat glands and hair follicles live and are teeming with blood vessels. When there is a deep cut, blood flows. Beneath the dermis is the skin's third and last layer, called the hypodermis, which is in contact with the body's bones and muscles and contains a lot of fat cells that help keep the body warm and insulated.

In all the years Professor Fehlinger studied pathology, he had never seen cadaverous skin tissue like he was examining. There was no identifiable separation of layers. The skin had solidified into one solid mass, which caused all the skin cells to die and all the blood vessels to break down. Professor Fehlinger next examined the internal organs taken from different cadavers, and they all had the same condition. They were all hard and stiff. Professor Fehlinger hypothesizes the internal organs died as a result of oxygen starvation, which is carried by blood, which stopped flowing when the blood vessels broke down as a result of the skin solidifying into one mass. But why? What was the cause that triggered the skin mutation?

When Professor Fehlinger finished, it was late into the night, he was tired, and his eyes were bloodshot red. Instead of returning to the Marriott hotel on Peach Street, where he and Liz had a suite provided to them by the CDC, Professor Fehlinger stayed in an office at the CDC that had a cot inside of it used for when scientists were working on something very important and wanted to stay near their work in case they had an epiphany. Liz returned to the hotel. She told her father she would bring fresh clothes in the morning. Scotty went home, which was a short distance from the CDC.

Professor Fehlinger slept lightly. It was how his mind worked. If he was confronted with a complex problem, he would sleep on it and give his thoughts free rein to go wherever they wanted. Professor Fehlinger's thoughts went back to when he was a student at Inuit Junior High in Anchorage, Alaska. It was where he first became interested in pathology. In Mrs. McCandle's class. He had to dissect a frog infected with worms. Since then, he had studied and dissected all sorts of species and developed a repertoire of knowledge where he could compare and contrast every repository of knowledge, he had accumulated over the years studying pathology, and there was nothing comparable to the Stone Man virus.

After a few hours of sleep, Professor Fehlinger climbed off the cot and returned to the lab. Professor Fehlinger accessed the lab computer. The computer's megabytes of knowledge had no comparable information.

Professor Fehlinger heard a knock at the door and looked up. It was Liz. She held up a McDonald's white breakfast bag with its golden arches emblem on the sides with one hand and a plastic hanger draped with clothes in her other hand.

Professor Fehlinger walked to the door and opened it. He stepped out. Although it was early in the morning, there were other scientists moving around quietly in their white smocks. Professor Fehlinger didn't want to distract them.

"Thank you," Professor Fehlinger said, taking the bag from Liz as they walked toward the cafeteria. They found a table and sat down. Professor Fehlinger opened the bag. A steaming cup of black coffee, a small carton of milk, a small carton of orange juice, two egg McMuffins, a small Danish, and three packets of sugar were inside.

"Have you eaten?" asked Professor Fehlinger, sipping the hot coffee.

"Yes," answered Liz, laying the hanger draped with clothes over the back of a chair. "Have you made any progress?"

"None at all," answered Professor Fehlinger.

"Rough night?"

"Yeah. How did you know?"

"Because whenever something is bothering you, you drink your coffee black."

"Am I that obvious?"

"Yes, Dad," Liz said, smiling, hoping to reduce her father's stress levels.

Professor Fehlinger didn't say anything as he bit into the egg McMuffin. He chewed the muffin slowly and meticulously, as if his mind were somewhere else.

"What's wrong, Dad?" asked Liz, reaching across the table and squeezing Professor Fehlinger's hand.

"I've been thinking about this thing all night, and I keep drawing blanks. I feel like I'm missing something."

"Mind if I join you?" asked Scotty. He was standing at the table, holding a cup of steaming hot coffee. He looked haggard.

"Sure, have a seat," Professor Fehlinger said, waving his hand at an empty seat next to Liz.

"Have you made any progress?" asked Scotty, sitting down.

"My daughter just asked me that same question, and I'll tell you the same thing I told her. None at all. I'm sorry, Scotty."

"Sorry for what?" asked Scotty.

"I know how important finding a cure for this thing is, and I know you were counting on me," answered Professor Fehlinger. "I feel like I'm missing something, but I can't put my finger on it."

"You don't have to apologize for giving your best shot. Besides, we're still in the race."

The table fell silent as both men sipped their coffee. Afterwards, Scotty said, "While you were examining the tissue samples, I got a call from the White House. The president wants a meeting to discuss options. He's going to declare the Stone Man virus an epidemic and declare a state of emergency. The army will be activated to quarantine Alaska and Antarctica. Russia and Canada are also coordinating their efforts with ours."

Professor Fehlinger nodded his head understandingly. "How bad is it?" Professor Fehlinger asked.

"It's really bad and getting worse. Deaths are in the hundreds and climbing. Confirmed cases of infection are in the thousands. The news media exposed a cover story about a new strain of flu that was initially put out by a hospital in the boondocks. People are scared and starting to panic and flee from areas of infection, which is why the president wants this meeting, and the army is being called out to maintain the quarantine areas."

"Jesus," Professor Fehlinger said almost to himself, which only emphasized the gravity they were all facing.

"I want you to accompany me to the White House."

"Me? Why me? I mean, I think I can be more of a help to you here, and besides, I'm sure there are other scientists better suited for that than me."

Scotty understood Professor Fehlinger's reluctance. He was aware of how his wife's research funding was abruptly cut. "There are a few reasons why I think you would be my best pick. You said so yourself. You haven't made any progress, and maybe a break is what you need. You're right. There are other scientists who could accompany me, but they wouldn't have the same effect as you. The CDC is a bureaucracy of the government, and I'm a political appointee. You're an independent and well-respected pathologist with no political agenda."

The table fell silent again as Professor Fehlinger counseled his own thoughts. Scotty's points were all valid. After a few seconds ticked by that seemed like minutes, he said, "I've come this far. I might as well go the distance. When do we leave?"

For an answer, Scotty peered over the top of his cup as he sipped more coffee. Professor Fehlinger had seen that look before and understood it as he reached for the clothes Liz brought.

Chapter 6

Most people would consider going to the White House to meet the president an honor and a privilege. However, Scotty, Professor Fehlinger, and Liz were unfazed. Each for their own reasons.

When Scotty was appointed director of the CDC, it was the first time he had met with the president. Although Scotty had the credentials for the appointment, graduating at the top of his class from Howard University with a master's in biology, he knew his appointment was a concession to the black community in Atlanta for getting out the black vote. On the president's desk in the Oval Office was a plaque that read, The Buck Stops Here. It was put there by President Eisenhower. Scotty knew from this president's temperament that he didn't feel that way. The current president felt that when something went wrong, it was everyone else's fault except his. The plaque on his desk should have read, The Buck Passes Here. Scotty knew the president had the same information as the CDC, if not more, and that he was in the direct line of fire of the president's ire.

In Liz's young life, she had heard about how politics had impacted her mother's career just when she was on the verge of proving a hypothesis she had dedicated her life to when the research funding she was depending on was abruptly cut. Liz hoped going to the White House would somehow help her understand why and perhaps provide a degree of closure.

Professor Fehlinger thought about how ironic life was. He was going to participate in the same political process by offering his expertise to the same office that ruined his wife's career. Professor

Fehlinger rationalized his participation by deciding that the needs of the many outweighed those of the few or the one.

The limousine stopped at the front entrance of the White House on Pennsylvania Avenue. They were met by a contingent of secret service agents. "My name is Sidney Williams, and I'm in charge of this detail. I need to see everyone's identification, please."

Scotty, Liz, and Professor Fehlinger showed their identification.

After examining them, Agent Williams said curtly, "Follow me."

They were encircled by the agents as they walked along the red-carpet walkway leading up to the White House. They passed a soldier in full military regalia standing at attention as they entered the White House. They walked along the clean, polished-to-shine white marble corridors. No one spoke as they passed people carrying briefcases or stacks of paper. They walked briskly, as if they were in a hurry. They stopped in front of a door. Agent Williams knocked twice in rapid succession. The door was opened by a sharply dressed woman wearing a dark blue power pants suit. She had long blonde hair that was pulled back tightly into a bun. She was very attractive, with an air of no sense about her. After acknowledging the other agents first, she said, "Please, come in."

Liz, Scotty, and Professor Fehlinger entered the room.

The room was large. A huge table dominated the room. Sitting at the table were some of the White House staff and cabinet members. Behind them sat their attendants with briefcases, stacks of paper, and laptop computers that they balanced on their knees or on top of their briefcases. Along the room walls were a bank of monitors. On top of the table were phones, more laptop computers, stacks of paper, pens, and pencils; tall crystal glasses; a crystal pitcher full of cold ice water; next to it was an ice tray; and a bowl of the president's favorite candy, M&M'S.

"Please, sit down," the woman said, motioning with her hand at three chairs on one side of the table. After they sat down, the woman retreated to a side door along with the other agents.

After a few minutes, the door opened almost magically, and in that instant, the aura in the room changed. Scotty, Professor Fehlinger, and Liz recognized the president immediately as a chair was pulled out at the head of the table for him to sit down. After he was comfortably seated, the president cleared his throat and said, "Good afternoon, ladies and gentlemen. I called this urgent meeting to discuss the pandemic sweeping across the world, mostly the northern hemisphere, dubbed the Stone Man virus because of its effect on the human body. Out of respect for their family members, I won't get into the details, as we all know what they are from various reported sources. Before I get into the details of this meeting, I want to take this opportunity to introduce everyone. At my immediate right is Vice President Earl Roberts; sitting next to him is Secretary of State Alvin Wright; next to him is Secretary of Defense Thomas Reed; sitting next to him is Secretary of Health and Human Services Latonya Robinson; and to my immediate left is White House Chief of Staff Edward Whitehead in charge of communications; sitting next to him is cabinet member Randolph Green in charge of coordinating policy; sitting next to him is Senior White House Counsel Ray Cagno; next to him is my adviser Frank Padro; next to him is Press Secretary Laverne Bradshaw; and lastly, sitting next to her is my National Security Adviser Jack Gunther. On the three wall monitors in front of you are Mozette Petway, my Ambassador to Canada; Frank Sutton, my Ambassador to Russia; and Margo Daye, Governor of Alaska." As the president made the introductions, they nodded their heads in acknowledgment.

Although Scotty knew they would never have made it on the White House grounds without identification, he introduced himself, Professor Fehlinger, and Liz. They followed protocol and nodded their heads in acknowledgment after being introduced.

The president began speaking again. "Now that everyone has been properly introduced, let's get down to business and why I called this meeting. I want answers, Scotty. What is going on with this virus? I hope you're not sleeping at the wheel." The accusation was obvious.

"Not at all, sir. As soon as the CDC became aware of how contagious this virus is, we've been working literally around the clock trying to develop a vaccine."

"And where are we on that?"

"So far, sir, we've been unsuccessful."

"I didn't appoint you as the director of the CDC to be unsuccessful. I appointed you because I was advised you were the best man for the job. Someone who could get things done over there." The room grew quiet at the president's chastisement.

Professor Fehlinger cleared his throat.

"You have something you would like to say?"

"Yes, I do sir, with all due respect," Professor Fehlinger said, unintimidated by the president's bullying tone.

"There are two reasons why the CDC is not caught up to speed with this thing. The first is that many years ago, the White House abruptly cut research funding across the board that would have gone into finding cures and vaccines for these sorts of outbreaks, and those funds were never restored. Secondly, this is a very complicated virus, unlike any other virus we have encountered before. I'm sure inevitably we'll develop a vaccine for this thing."

"Inevitably? What does that mean? Because if it means time, that's something we don't have. Last I checked we're losing."

"Not really, Mr. President, sir," Liz said, jumping in. The dig her father made about the White House cutting research funds and not restoring them didn't escape her. Now it was her turn. "These outbreaks have always occurred in our history. I won't give you a litany of outbreaks we've had in human history except for the last one, which was the bubonic plague. The disease killed half of Europe's population before a vaccine was developed. It took time, and that was with the full weight of the government funding research for a vaccine. As my father, Professor Fehlinger, stated, our efforts to fast-track a vaccine for these sorts of outbreaks are hampered by the White House's lack of research funding."

The president could feel the political heat building and didn't like it as he said, "How long did it take to develop a vaccine for the bubonic plague?"

"Technically, whenever there's an outbreak, the struggle for survival starts immediately. Mr. Scott and my father are working feverishly to develop a vaccine. It took two years to develop a vaccine for the bubonic plague. I think with increased research funding that should never have been cut in the first place, a vaccine, perhaps, could be developed in less than two years."

Professor Fehlinger was proud of his daughter. She was not intimidated by the president and had spoken truth to power.

"Suggestions?" asked the president, looking around the room.

Latonya Robinson was the Secretary of Health and Human Services in 2016, when the then President of the United States abruptly cut all research funding on the premise that global warming was a hoax promoted by scientists with a socialist agenda. She had gone along with the cuts to safeguard her job and career. She should have spoken up then, but she was too intimidated by the then president, and now his son is following in his father's footsteps because it worked then. She would not be cowed this time. Liz Fehlinger gave her the courage to find her voice. "We should immediately double federal funding for research to develop a vaccine and raise the exposure level from an epidemic to a pandemic."

Margo Daye, Governor of Alaska, said, "I recommend Alaska be declared a state of emergency so we can access some of those federal funds. The first influx of funds will be initially used on purchasing hazmat suits for our first responders."

Laverne Bradshaw, White House Press Secretary, said, "You should do a news conference. It'll make you look strong and give the American people confidence you're on top of this thing."

Thomas Reed, Secretary of Defense, said, "Alaska, where the virus seems to be hitting the hardest, should be put under Martial Law, and a mandatory curfew established from noon until twelve o'clock sun rise for nonessential workers."

Ray Cagno, Senior White House Counselor, was shaking his head negatively. "No, we can't do that. The Armstead Act prohibits the military from engaging in military action on American soil."

The president dipped his hand in the bowl of M&M'S and began eating the chocolate-covered coated peanut candy, tossing them into his mouth one by one as he listened to the suggestions.

"We don't have to use the military for that. The governor can call out the national guard," Thomas said.

"Okay," the president said, holding up a chocolate-smudged hand. Someone behind him gave him a napkin. "I heard enough to know what I am going to do." He held up the napkin, and someone took it away. The president looked at his watch and said, "Laverne, I want you to schedule a news conference for five o'clock. I'm going to announce an increase in federal funding for research into finding a vaccine for this virus. I'm going to declare a state of emergency for Alaska, which will qualify the state to receive federal funding. I'm going to order the CDC to raise the exposure level from an epidemic to a pandemic. This will be our new policy. Ed and Randolph, I want you two to work together on this so this will be a smooth rollout. Margo, I need you to make a request for national guard deployment so I can send it. I want you to call for a curfew from noon until twelve o'clock sun rise for essential workers. It'll look better coming from you," the president said, looking at the monitor.

"You have my full support," Margo said.

"The national guard will do the enforcing. The army will act as support only. Will that take care of the Armstead Act, Ray?"

"It will, Mr. President."

"Scotty, I want you to attend the press conference with me."

"Yes, Mr. President."

"If there's nothing else, I declare this meeting adjourned so I can start preparing for the news conference," the president said, and with that, the meeting was over. Everyone remained seated until the president and his vice president were out of the room. Once they were gone, everyone else left the room according to their job titles, which made Scotty, Professor Fehlinger, and Liz last.

As soon as they stepped out of the room, the contingent of secret service agents appeared almost magically. "Please, follow me," Williams said.

They were escorted to the White House press waiting room. The floor was covered with wall-to-wall blue carpet with a huge United States insignia recognized all over the world in the center. Along the walls were red leather couches and recliner chairs. The center of the room was dominated by a mahogany table laden with a variety of magazines and newspapers. The room had a huge television screen mounted on the west wall. At the back of the room was a small hallway that led to the bathroom.

"So, this is where the other half live when they're in front of the camera," Professor Fehlinger said, sitting down on one of the couches after the contingent of secret service agents disappeared.

"It's annoying how they keep appearing and disappearing. What do they think? We're going to steal something," Liz complained, sitting next to her father.

"It's protocol. I found it annoying too when I first came to the White House for my appointment," Scotty said, sitting in a recliner across from them.

"Have you thought about what you're going to say?" asked Professor Fehlinger.

"I'll echo whatever the president speaks on, and anything beyond that, I'll stick with the truth. Can't go wrong there."

Scotty opened up his brief case and went over his summary. Occasionally, he looked up to see Liz watching him. He smiled at her and continued reading.

Professor Fehlinger saw the looks. He didn't say anything. Liz was an intelligent, grown, and single woman with an independent mind. Professor Fehlinger picked a copy of the AMA off the mahogany table and started reading. Liz picked up a copy of Seventeen and started reading. The magazine was full of the latest teen trends, which allowed Liz to better understand and relate to her young students. No one paid attention to the time until the room door opened, and Laverne Bradshaw entered.

"We'll be ready to go in a few minutes," she said, looking at Scotty.

The president, his vice president, and Latonya Robinson entered the room with a contingent of secret service agents.

"Ready," the president said to Scotty.

"Ready, sir," Scotty said, standing up.

"Let's go then," the president said.

After they had gone, Professor Fehlinger picked up the television remote and clicked on the television. Professor Fehlinger and Liz watched as the president, his vice president, Laverne Bradshaw, and Scotty stepped out from behind a thick purple curtain. The president stepped up to the lectern. The vice president, Laverne Bradshaw, Latonya Robinson, and Scotty stood behind him. The contingent of secret service agents flanked them and discreetly posted up around the room, where they had the advantage of seeing everyone.

"Ladies and gentlemen of the press, and my fellow Americans," the president began. "As you know, the northern hemisphere, including Russia, the Antarctica, Canada, and parts of the United States, which include Alaska, are being ravaged by a deadly virus that has a high mortality rate. By executive order, I am raising the exposure level from an epidemic to a pandemic. I've declared a state of emergency in Alaska, which will allow for the assistance of federal funding for first responders and whatever else that state may need as the governor requests it from the White House. I have also doubled federal funding for research on developing a vaccine for this terrible virus. There will be a complete quarantine of the State of Alaska's border to contain the virus. I ask the citizens of the great state of Alaska to be patient. Those of you who seek to take advantage of this pandemic by price gouging, looting, rioting, and any other criminal activity will be prosecuted to the fullest extent of the law. Whatever questions you have, please refer them to my press secretary, Laverne Bradshaw, and Mr. James Scott, who is the director of the CDC." The president paused and turned to Laverne and Scotty, who nodded their heads before continuing, "I would love to stay and answer your questions, but I have a very important meeting to attend." As the president turned to leave, the questions started.

"How much money is the White House allocating to combat the virus?"

"Will your administration establish a task force?"

"Where did the virus come from?"

"Why is the virus concentrated in the northern hemisphere?"

"How long will the quarantine remain in effect?"

"What's the body count so far of the people who have died from the virus?"

The president ignored the questions as he disappeared behind the curtain along with his vice president.

Laverne Bradshaw stepped up to the lectern and began answering the questions. “With regard to the first question, the administration will initially be doubling research funding, and more funds will be available as needed. As to the second question, at this time, no task force is being considered. The president has full confidence in the CDC under the leadership of its current director, Mr. James Scott. No one knows where this virus originated. We have some of our best scientific minds looking into that as I speak. Nor do we know why the virus seems to be concentrated in the northern hemisphere or along our northern states. If you turn to the monitor on the west wall, I’ll let Margo Daye, Governor of Alaska, speak to the issue of how long the quarantine will be in effect.”

In unison, they turned to the monitor. “I’m sure everyone can understand why I’m appearing via satellite and why I’m not there at the White House. The quarantine will remain in effect as long as it’s necessary. The quarantine will start at noon and last until sunset for nonessential workers. I’m also deploying the national guards to enforce compliance with the quarantine and curfew with assistance from the army, which will only be providing technical support.”

“What is the death toll so far, governor?” asked one of the reporters.

“Unfortunately, those figures are constantly changing as more people succumb to the virus.”

“How is the virus spreading?” asked another reporter.

“I’m not a doctor or scientist, so I’ll defer that question for Mr. Scott to answer.”

Scotty stepped up to the lectern.

“We don’t know that yet,” he said.

“Does this virus have a life span?”

“We’re not sure.”

“Is there any particular age group being infected disproportionately?”

Scotty looked at his notes and said, “Although people are being infected indiscriminately, there seems to be just a slight uptick in the adult population, which may be contributed to younger people ranging from eighteen to thirty-two years of age having a stronger immune system."

"What is the best way we can protect ourselves from getting this virus?" asked a female reporter sitting up front.

Scotty looked straight at her. It was one of those questions he dreaded answering because there was no clear answer beyond wearing the standard face mask, which is what he suggested. "This is a very complicated virus."

"Complicated how?"

"What was that?" asked Scotty, peering in the back.

A reporter stood up from the back row of reporters and asked the question again. "You said the virus was complicated. Complicated how?"

"There are all sorts of different viruses that affect people differently, and we just don't know enough about this virus yet to make any definite conclusions," answered Scotty.

"When will you know? Has the CDC started work on a vaccine? Is there any one person leading the research on developing a vaccine? Russia and Canada have closed their borders. Could the virus have come from either of those two countries?"

The questions kept coming, and so did the I-don't-knows. When Scotty finally returned to the press waiting room, he was tired and sweaty from the glaring lights.

"That was brutal," Professor Fehlinger said as Scotty flopped down in a recliner chair. Liz brought him a glass of water. "Thank you," he said. After drinking the water, Scotty said, "Not having the answers to their questions was the worst part."

"According to all reports, Alaska is where the virus is infecting the most people. I think we should go there," Professor Fehlinger said.

"We may be able to work backwards to the source," Liz said.

"Those are very good suggestions," Scotty said, mulling their ideas over.

"How soon can you get us there?" asked Professor Fehlinger.

"It'll take about an hour to refuel the plane and another six or seven hours to reach Alaska, depending on weather conditions," answered Scotty.

"So... are we going?" asked Liz hesitantly.

"As soon as I can get to a phone and make the arrangements."

Chapter 7

Alaska is from the names "Aleutalaska" or "Eskimo alashak." Both names mean Mainland. Alaska's nickname is "The Last Frontier." Its capital is Juneau. Alaska became the forty-ninth state of the United States on January 31, 1959, with its motto "North to the Future."

Alaska has ten cities, beginning with Anchorage, Fairbank, Juneau, Sitka, Wasilla, Ketchikan, Kenai, Palmer, Kodiak, and Bethel. Anchorage, by far, is Alaska's biggest city. It also had the highest rate of infection and death from the virus.

These thoughts raced through Professor Fehlinger's mind as the plane flew into Alaska from the north east. Professor Fehlinger looked across the aisle.

Scotty was busy looking over papers, writing, and talking on his cell phone. Professor Fehlinger looked at Liz. She was sitting behind Scotty, looking out the plane window, absorbed in her thoughts. She must have felt his gaze because she looked over at her father looking at her, smiled warmly, and turned back, looking out the window. Professor Fehlinger wondered what she was thinking about. It was the first time they had been back to Alaska since Elizabeth had died.

"Can we fly low enough to see the landscape?" asked Liz.

She was talking to Scotty, but he didn't respond. Liz tapped him on the shoulder.

"Yes," he said, turning around in his seat.

"I was asking you if the plane can fly low enough to see the landscape."

"I'm sorry. I didn't hear you. I was preoccupied with keeping up with the latest status reports and making arrangements so our visit to the hospital would go smoothly. I'll check with the pilot."

Scotty knocked on the pilot cabin door. After a few seconds, the door opened.

"Yes?" the pilot said.

"Can we fly low enough to look over the landscape?"

The pilot thought for a few seconds and then said, "Sure. We can do that. The sky is clear, and there's no air traffic to worry about."

"Thank you," Scotty said. He returned to his seat and told Liz what the pilot said. They could feel the plane starting to descend slowly.

Professor Fehlinger looked at Liz, but she continued looking out the window. Professor Fehlinger turned in his seat and looked out of his window, wondering what his daughter was so preoccupied with.

At some point, Professor Fehlinger must have dozed off because the pilot's announcement to fasten your seat belt for landing over the intercom awakened him. Professor Fehlinger fastened his seat belt. He looked over at Liz and Scotty. They were fastening their seat belts.

When the plane landed, it was dusk. Instead of being met by a limousine, they were met by a blue and white four-door national guard car escorted by an army jeep with a mounted machine gun.

"My name is Sergeant P. Bosley, and my orders are to escort your entourage to the general hospital," he said to no one in particular, not knowing who was in charge of the entourage.

"My name is James Scott. I'm the director of the CDC. This is Professor Fehlinger and Ms. Fehlinger."

Sergeant Bosley shook each of their hands, and after the introductions were made, he said, "Follow me, please."

Sergeant Bosley slid into the car driver seat. Scotty sat next to him. Professor Fehlinger and Liz sat in the back.

"Is this really necessary?" Scotty said as they followed the army jeep.

"I'm afraid so. The people are really scared, and when there is fear and panic, the worst and best come out of the people," Sergeant Bosley explained.

As they drove along, the streets were eerily empty because of the mandatory curfew. The only people out were national guardsmen supported by the army. Trash and other debris were strewn along the streets.

When they reached the hospital, it looked like a militarized zone. There were national guardsmen posted around the hospital. A huge tent was erected at the front entrance.

"Sergeant Bosley, requesting permission to enter sir!"

"Request granted," came a curt response.

They entered the tent. Inside was a table with papers, stacks of folders, phones, laptop computers, and maps on top. The tent also had a microwave, a huge ice cooler, a refrigerator, cots along the sides, foot lockers, and a short-wave radio. There were four chairs at the table that formed an inner circle, and four more chairs around the tent that formed an outer circle. Two men sat at a table in an inner circle across from each other. The color of their uniforms and insignia identified both men. One man was a lieutenant in the national guard, and the other man was his counterpart in the army.

"Lieutenant Ryan, sir! This is Mr. James Scott, Director of the CDC, Professor Fehlinger, and Ms. Fehlinger," Sergeant Bosley said, after first saluting the two men stiffly.

"At ease, Sergeant," the man, Sergeant Bosley addressed as Lieutenant Ryan said, coming around the table with his hand stretched out. "I'll take it from here. Sergeant."

"Yes, Sir!" Sergeant Bosley said.

"I was told to expect you and your associates," Lieutenant Ryan said, shaking Scotty's hand. "This is Lieutenant Norris. He's with the army."

After a round of shaking hands, Lieutenant Ryan said, "I suspect you want to get started as soon as possible."

"Please," Scotty said.

Lieutenant Ryan walked over to one of the foot lockers and opened it. "All of you will need to put these on," he said, passing out hazmat suits and foot booties.

After they were all wearing hazmat suits and the foot booties, Lieutenant Ryan said, "Follow me and be careful."

Entering the hospital was like stepping into the most poverty-stricken third-world country. The floor was covered with blood, vomit, phlegm, and all sorts of paper debris, and only God knew what else. There were patients everywhere. Sitting, standing, and lying down, some were propped up against the hospital walls, and they were all infected with the virus. Some of the patients appeared to be dead. There were children, too. That was the worst part. Armed national guardsmen patrolled the hallways and were posted at doors. They were wearing hazmat suits too, along with the doctors and nurses.

They took the elevator to the second floor, where the hospital administrator's office was located. Inside the office waiting for their arrival were Douglas White and Teresa Lemons. Scotty knew them both. Nonetheless, Lieutenant Ryan made the introductions.

"The air in this office is purified. You can take off your head gear," Douglas said.

"It's been a long time since I last saw the both of you," Scotty said, taking off the respirator.

"Yes, it has been, and I'm sorry it has to be under these dire conditions," Douglas said.

"Does or can this purified air kill the virus?" asked Professor Fehlinger.

"His field of study is pathology," Scotty said.

Douglas nodded his head understandingly and said, "Then you know viruses can't incubate in purified air. With this particular virus, though, I just don't know. I've never seen anything like it before. The purified air is just another added layer of protection, as I assume none of you have been infected, which raises another question. How are these infections being transmitted? So far, the only common denominator seems to be that once a person becomes infected, that person can transmit the virus to someone else. So we know the virus is contagious as hell. What we don't know is the source."

"That's why we're here. To try and find out more about this virus," Scotty said. "Have the arrangements I requested been made?"

"Yes," Teresa said. "I don't understand the purpose, though. I faxed the CDC all the information we have about the virus, along with its different stages of progression for its victims."

"I asked Scotty to make those arrangements," Professor Fehlinger said. "Scotty has shared with me all the information that was collected about the virus for my input. However, I'm sure you can appreciate the value of examining a patient firsthand as opposed to reading medical reports."

"I agree with you, Professor. It wasn't clear to me why Scotty asked for those arrangements until now."

"If everyone is ready, we can start now," Douglas said.

"Let's," Professor Fehlinger said. "If we're going to beat this thing, time is really important."

They all put on their hazmat suits and left the office with Lieutenant Ryan leading the way. When they stepped out of the elevator on the first floor and started walking toward intake, a man suddenly lunged at Liz, knocking on her respirator.

"Help me! Please me!" the man screamed, his voice just above a whisper, sounding hoarse and dry. One of the national guardsmen patrolling the hallway slammed the butt of his rifle into the back of the man's head. The man collapsed to the floor in a heap.

"Are you okay?" asked Professor Fehlinger, concerned, touching Liz's arm as she adjusted the suit's respirator.

"I'm okay," Liz answered as she knelt to check on the man. The blow had not broken the man's skin, and there was no bleeding. Under normal circumstances, she would have insisted the man be checked for a concussion, but the circumstances were far from normal, and they were there on a far more important mission, so she kept her peace. She glared at the national guardsman angrily to express her dissatisfaction. The guardsman was unfazed as he looked on stoically.

The intake area of the hospital was where the very first patients suffering from the virus were taken. The first patient they saw was a man named Frank Austin. Douglas explained to Frank who Professor Fehlinger was, and he sought his permission to have Professor Fehlinger examine him. Frank gave his consent. He would have agreed to anything. He was really scared. He saw the effects of the virus on other people and knew what was happening to him.

"What made you come to the hospital?" asked Professor Fehlinger while checking Frank's pulse.

"I started coughing and feeling miserable yesterday, and all the news reports said that's how the virus starts, so I came to the hospital," Frank answered in between bouts of coughs.

"Have you been around anyone else coughing?" asked Professor Fehlinger, now checking Frank's lymph nodes along the sides of his neck.

"My wife and two kids. They're here somewhere. They have the same coughing spell as I do."

Professor Fehlinger looked at Scotty. He understood and walked out of the room. Professor Fehlinger checked Frank's temperature. It was one hundred. "Does this or that hurt?" asked Professor Fehlinger, manipulating Frank's joints. His elbows, wrists, fingers, knees, and toes.

"Everything hurts," answered Frank, grimacing in pain with each manipulation.

After the examination was over, Professor Fehlinger, Douglas, Teresa, and Liz were escorted to a second patient by Lieutenant Ryan. Just before they entered the room, Scotty stopped them at the door and said to Professor Fehlinger, "I checked on the wife and kids. Apparently, the kids were sneaking out of the house to go play with other kids who are all now infected along with their parents, and Professor, his wife died just a few minutes ago."

Professor Fehlinger was disgusted by the information just as Lieutenant Ryan opened the room door. The second patient was a woman. Her eyes were open, but vacant. Next to her bed was a stand holding a small plastic bag filled with morphine that dripped through a tube into her arm to alleviate her pain. At the foot of the bed was the woman's chart. Douglas scanned the chart and then passed it to Professor Fehlinger. After reading the woman's chart, Professor Fehlinger walked around to the head of the woman's bed and said to Douglas, "Do you have a pen light?"

Douglas searched through his pockets and then said, "No."

"Here," Lieutenant Ryan said, "take mine."

"Thank you," Professor Fehlinger said, taking the pen light. Professor Fehlinger held one hand over the woman's face and, using his fingers, manipulated the woman's eyelid open while using his other hand to point the pen light into the woman's eyes. There was no reflection. There was no dilation. There was nothing.

"Can she see us?" asked Liz.

"No," answered Professor Fehlinger. "In order for our eyes to see the entire retina, which is the pupil, have to dilate. I suspect that with every second that goes by, her internal organs are solidifying. I'm sorry, but I have to do this." Professor Fehlinger manipulated the woman's joints as he had done with the first patient, but this time he was met with more resistance. When Professor Fehlinger finished, there were tears streaming down the woman's cheeks. Professor Fehlinger apologized again for what he had to do and then turned to Douglas and said, "I want her drips increased. Is that something you can do?"

"Sure. I'll take care of it. Ms. Fehlinger, as I understand it, you're an anthropologist, which is a lot different than looking at live people, so I want to warn you that the next patient we see is in the final stages of the virus and his appearance." Douglas paused for a few seconds, trying to find the right words before continuing, "Like if you can imagine what happens to a piece of fruit left out in the sun all day."

"You don't have to do this," Scotty said.

"I'll be okay," Liz said.

Liz thought she was prepared. She had seen the remains of mummified ancient people and thought this would be no different. However, Liz had never been up close and personal with death. Liz gasped at the figure that was once a human being. She turned her head away from Scott's shoulder.

The patient was a man. He had transmogrified into a stone mummy. His skin was a grayish color, wrinkled, very dry, and hard. His head had shrunk like a raisin. Professor Fehlinger had seen death before. He began the grim task of probing the man's body. The man's joints were so stiff and hard that it was impossible to manipulate them without causing breakage. The man's body had become twisted in death. Like a pretzel.

When they returned to the administrator's office, the first few seconds were spent in silence as they took off their respirators. Douglas spoke first. "What we just saw is multiplied by the hundreds all over this hospital and other hospitals as well, and by the thousands as the general condition of the state with people infected by the virus."

"That poor man," Liz said, sitting next to Scotty. It didn't escape Professor Fehlinger's attention; they were holding hands.

"The patients we just saw—I want to read their case files. Is there some way I can do that alone? I need to think," Professor Fehlinger said.

"Sure," Douglas said. "You can use my inner office. There's a desk computer on the desk. Just punch in the patient's name and their admission number, and their file should come up. Our services, whatever we can offer, will be better used out on the floor," Douglas said, looking around the office. When there were no objections, he continued, "When you're finished, you can either join us or press that red button on the computer's keypad, and it'll connect you to the intercom system."

"Thank you," Professor Fehlinger said.

"No. Thank you for whatever light you can shed on this thing," Douglas said as he put back on his respirator.

When the office was empty, Professor Fehlinger took off the bulky hazmat suit and went into the inner office. Professor Fehlinger sat behind a desk and accessed the computer, punching up the patients he had examined. The reading went quickly. There was no new information about the pathology of the virus. The first reported case of the virus was in 2006. No one understood it then or how contagious the virus was. The second reported case was eleven years later, in 2017. The crewmen of the ship named Martha were all infected with the virus and had to be towed into port by another ship named the North Star. Professor Fehlinger cross-referenced the names of everyone who had come into contact with someone infected with the virus. There was a pattern. The pattern that emerged had gaping holes, but it was more than what he had started with. Professor Fehlinger was able to trace the progression of the virus from John Wainwright to Martha's crewmen to the North Star. Once Martha and the North Star were brought into port, the virus infected the health care workers at the hospital before they knew the danger they were in. The North Star crewmen were allowed to return to their communities, back to their families and friends, and back to jobs where the virus spread exponentially. However, what the pattern didn't show or explain was why the virus lay dormant for those eleven years, and why were the northern parts of the hemisphere and the United States more impacted than the southern parts?

There was another group of scientists from the CDC at the WRC in Antarctica working with the Russians to study the virus. Professor Fehlinger decided he needed to go there. Professor Fehlinger pressed the red button.

Chapter 8

Back at the White House, in the famously secure situation room located in the west wing, sat the president, his vice president, Secretary of State Alvin Wright, Secretary of Defense Thomas Reed, National Security Adviser Jack Gunter, Secretary of Homeland Security Raymond Smith, Joint Chief of Staff General Vernon Collins, and Latonya Robinson. The meeting was called to discuss contingency plans to stop the spread of the virus and to present options to the president.

"The two things I'm most concerned about are stopping or at least containing the spread of this virus and the security of this country. So, I want those two issues addressed first before we get into the weeds on other issues," the president said, starting the meeting. "Latonya, you're up first. What do you have for me?"

"It's really too early to tell, but all indications are that the virus is either slowing down or it's reached its peak."

"Why the uncertainty?" asked the vice president.

"Because we just don't know enough about the virus to reach any definitive conclusions."

"Jack?"

"Activating the national guard, Mr. President, as early as you did was a good decision. It allowed us to clap down early. However, in those parts of the country infected the most with the virus, we've had to use a heavy hand to stop looting, maintain curfews, and keep people within their respective local borders. I'm afraid, sir, if the virus spikes, you may have to declare martial law."

"Shit!" the president said. No one was surprised by the invective. They knew before he was elected that he was inclined to use profane language to make or emphasize a point.

"What about the Russians?" asked the vice president.

"There infection rates are double ours. Their mortality rates are also double ours. They've erected a fence fifty miles inward from Alaska's borders to stop their citizens from crossing over and ours from entering their country. They have a group of scientists working alongside ours at a WRC in Antarctica," Thomas said.

"What's our move if this virus gets worse?" asked the president.

There was silence in the room.

Jack looked at the president and vice president. The president and vice president exchanged glances. The president cleared his throat and said, "The vice president and I have another very important meeting scheduled. We'll leave with instructions to continue the meeting."

After the president and vice president left the room and the door was securely closed, General Collins said, "I want to remind everyone that anything said in this meeting is highly classified." He paused long enough for emphasis before continuing, "We've been conducting experiments on the virus for possible military applications."

"Jesus!" Latonya said, interrupting him. "You people never learn."

Thomas held up his hand and said, "Please. Let him finish."

"What we found is that under intense heat, the virus spreads rapidly and exponentially. The opposite effect happens when the virus is exposed to intense cold. If we have absolutely no other alternative, we can create an environment that will neutralize the virus."

"What do you mean by creating an environment that will neutralize the virus?" asked Latonya suspiciously.

"We can explode a bomb over the designated area that will suck the oxygen out of the air, which will cause the temperature to drop."

"What's the caveat?" asked Thomas.

"The temperature will drop thirty-two degrees below Celsius."

"English Vernon," Thomas pressed.

"I see someone wasn't paying attention in their science class. Thirty-two degrees below Celsius is the temperature water freezes at," Jack said.

"And anything else that's alive," Latonya said.

"I'm afraid so," General Collins said.

"You can't be serious!" Latonya said. "You're suggesting the United States government drop a fucking bomb that will kill thousands, if not millions, of its citizens!"

"Before you condemn me with your self-righteous moral indignation, let's look at the alternative," General Collins said harshly. "Thousands have already died, with more deaths predicted when you include the number of people infected with the virus. The virus has a 100 percent mortality rate, which means millions of people will die unnecessarily if we don't stop the spread of this virus. The economic fallout of so many people dying will collapse the economy and make the 1930s depression look like an economic boom. The country will go to hell in a hand basket. Is that what you want?"

"Of course not, and you know that. There has to be another way. It just has to be!"

"I'm listening," General Collins said sarcastically.

"I don't know, but dropping a bomb can't be the only answer," Latonya said, not wanting to give up.

"Believe me," General Collins said wearily. "I know what I'm proposing. I've thought about this over and over from every angle you can possibly think of. There is no other way."

"Does the president know?"

"Right now we're just talking, but if any of this were to leak out before its time, it's called plausible denial."

"The vice president too?"

"What do you think?"

Latonya suddenly understood the meaning of the president and vice president's impromptu meeting. Their absence was their shield of protection. "This is too big of a decision for one person to make. We should take a vote."

"I agree," Alvin said.

Chapter 9

Scotty, Professor Fehlinger, and Liz flew to Anchorage. They flew from their military helicopter to Antarctica and landed at the WRC. It was still early in the afternoon when they arrived. They were met by Dr. Davalloo and two other men she introduced as William Fiske and Shaun Rossi. Scotty, Professor Fehlinger, and Liz were surprised they were not wearing hazmat suits.

"No suits," Scotty said, voicing all of their thoughts.

"I'll explain later," Dr. Davalloo said. "Once we get inside."

William and Shaun helped carry Liz's suitcases as they made their way to the WRC. Their heavy, thick-hooded white coats and several layers of clothing protected them from the freezing cold. The hazmat suit Scotty, Professor Fehlinger, and Liz were wearing protected them from infections but did very little to protect them from the cold. They were relieved to get inside the warmth of the WRC.

"When I first arrived here, I found the dead bodies of the workers. None of the bodies had any signs of external injuries that would account for their deaths so I rationalized that whatever caused their deaths had to be some kind of virus, and since the host was dead, there was no way for the virus to survive or multiply. I put my hypothesis to the test by taking off my hazmat suit. Nothing happened. I went outside. Nothing happened. I was right. After a few days, I stopped wearing my hazmat suit, and when the others arrived, so did everyone else."

"Isn't that dangerous?" Scotty said. "All the data we have indicates the virus appears and then disappears before reappearing for reasons that are still unknown."

"Of course it was. However, we are scientists. We can't allow uncertainty to stand in the way of observation and experimentation with the phenomenon. When you explained to me the pattern Professor Fehlinger found, I realized the only people I had come into contact with were the other group of scientists, and the people who worked here, and they were already dead from the virus, and that was weeks ago. So, in essence, my team and I have been in self-imposed quarantine. The risk I took was really minimal."

"You're a gutsy lady," Professor Fehlinger said, speaking for the first time. "Practical too," and with that, Professor Fehlinger took off his hazmat suit. Scotty and Liz followed in tandem.

Dr. Davalloo gave them a tour of the WRC and where they would be sleeping. After they unpacked, they joined the other scientists sitting at the huge table in the hub of the WRC. Dr. Davalloo was the point person for investigating the virus.

"The people sitting here are experts in their field of study," she said. "You've already met the two gentlemen to my right. William is an optometrist, and Shaun is an epidermis. Next to him is Hal Roberts. He's an epithemist. Next to Hal is Gretchen Lawrence, and next to Gretchen is Barbara Jean. Gretchen is a biochemist, and Barbara is a biologist. Last, but not least, is Derrick Mayhue. He's an orthopedist. To my left are our Russian counterparts. Sergei Arilov, Ivanov Alexander, Legkov Borish, Vladimir Malkov, Nikita Kriukov, Lydia Gusakova, and Nikolay Ivanovich. Ladies and gentlemen, our latest arrivals are Mr. James Scott, Director of the CDC, and a cell biologist, Professor Fehlinger, who is a renowned pathologist, and Ms. Fehlinger, who is an anthropologist."

After the introductions were made and everyone exchanged courtesy nods, Dr. Davalloo continued, "I crossed-referenced this virus with other known viruses comparable to what we are facing, and there are non-comparable viruses. All my references came back negative. I checked for similarities. Those came back negative, too. This virus is unique. Generally, a particular virus will only disrupt the part of the body that's infected. However, this virus attacks the whole body and breaks down all of the body's functions."

"Dr. Davalloo. If I may," Barbara said.

"Of course," Dr. Davalloo said.

"Dr. Davalloo is right in describing the virus's uniqueness. Although the virus attacks the whole body system, it does not do so all at once. The first symptom is a nagging cough, which means the virus attacks the respiratory system first by hardening the diaphragm, making it difficult for the lungs to expand."

"So, the virus attacks the soft tissues of the body first, and then the bones," Nikolay said, in English with a thick Russian accent.

"That's what the data indicates," Barbara said.

"I just left the general hospital in Anchorage, where I stayed for a few days. While I was there through an arrangement, I had the opportunity to examine patients who were infected with the virus in different stages, and what I found was that the bones were the last stage of infection before death," Professor Fehlinger said.

"That makes perfect sense," Derrick said. "Our bones are already hard because they are made of calcium and phosphate, which are the same minerals in actual rock. The virus accelerates the body's soft tissue into calcium and phosphate because of the presence of these minerals already in the bone."

Although Liz was an anthropologist, she was able to keep up. She found the depth of knowledge about the human body being articulated at the table in such detail fascinating! Liz sat silent, writing furiously. Everyone around the table was taking notes.

It was late in the afternoon when they broke for lunch. After eating lunch, everyone returned to the table. When night arrived, everyone was mentally exhausted. The plane ride from Washington, D.C., then taking a helicopter from Anchorage to the WRC, and the high-level complicated discussions took their toll. When Scotty, Professor Fehlinger, and Liz went to bed, they slept the sleep of the damn.

When Liz awoke, it was a little past noon. She walked to the WRC cafeteria and asked the cook what was for breakfast.

"Breakfast?" the cook said raising an eyebrow. "Breakfast was hours ago. If you still want it, I can whip up almost anything."

"Thank you," Liz said, feeling a little sheepish. "I'll have two scrambled eggs, two pieces of toast, and straight black coffee." Liz knew the caffeine would give her a jolt.

"Coming up," the cook said, and she began scrambling the eggs.

There was a lone man sitting at the table, sipping a steaming cup of coffee, and reading some papers. Liz sat at the table and said, "I hope I'm not disturbing you."

"No, you're not." He stopped reading momentarily, focused his attention on Liz, and said, "I saw you yesterday when you came in. My name is Skip Wilder. Everyone calls me Skip."

"My name is Elizabeth Fehlinger. Everyone calls me Liz."

"If you're looking for your associates, they're in the lab with the other white coats."

"White coats?"

"Yeah. That's what we call the scientist," Skip explained.

"The older gentleman wearing the glasses is my father. He's a professor. His name is Sterling Fehlinger, and the other gentleman with him is the director of the CDC. His name is James Scotty."

"No offense."

"None taken."

"So, you're not a…"

"White coat," Liz said, finishing the sentence.

"Busted," Skip said, not wanting to deny what he was thinking.

"No, I'm not. I'm an anthropologist. What about you?"

"I'm the lead meteorologist."

The table fell silent as the cook set Liz food in front of her, along with a cup of steaming black coffee.

"If you don't mind me asking what you are doing," Liz said as she sprinkled her eggs with salt and pepper.

"I'm going over my notes forecasting the weather so I can send them out to various radio and television stations in the area."

"Do you like it?"

"I love it. I love being out in the fresh air. I'm originally from the coal mines of Kentucky. I come from a long line of coal miners, starting with my great-grandfather. I was expected to follow in their footsteps, but I chose college and meteorology. I tried coal mining and absolutely hated it. I felt like the canary."

Something clicked. "What did you say?" asked Liz, interrupting Skip. The cup of coffee is inches from her lips.

"Which part?" answered Skip, a little confused.

"About the canary."

"I wasn't finished before you stopped me. I was going to say I felt like the canary they keep down in the mines. Trapped."

"They were used to detect gas?"

"Back then, yes, but not now," Skip said, looking at Liz. He could tell there was something on her mind.

"I gotta go," Liz said hurriedly, without finishing her coffee or meal. She returned to her sleeping quarters, where she layered up and went outside. It was quiet. Too quiet. Liz looked around in every direction, as far as her eyes would allow her to focus. The landscape was completely blank. Devoid of any life. Liz walked around the perimeter of the WRC, expanding her walks with each circuit, until she returned to the WRC too hyped to feel exhaustion about what she had not found.

As soon as Liz was out of her layered clothing, she walked to the hub of the WRC, where the other scientists were gathered around the table. Liz sat down just as Dr. Davalloo was saying, "Unique as this virus is, it's still a parasite at its basic level."

"Excuse me. Can I say something?" Liz said.

"Sure," Dr. Davalloo said, a little annoyed at being interrupted.

"What animals are common to this area?" asked Liz.

"Polar bears, snow rabbits, reindeer, and seals are just a few of the animals that live in this area," Ivanov said.

"When was the last time anyone had seen any animals?" asked Liz.

The hub was silent. Then, Dr. Davalloo said, "I've been here longer than anyone, and I haven't seen any animals."

"When we were flying in from Washington, D.C., I didn't see a lot of animals, and the further north we flew, the animals seemed to disappear. I took a walk around the perimeter, and I didn't see any animals, so I kept walking about a quarter of a mile out, and I still didn't see any animals. No tracks, droppings, carcasses—nothing. I'm not a scientist, but as an anthropologist, I find that very odd."

At first, no one said anything as they thought about the question.

"When I first arrived here, I made those same observations, but I didn't think anything about them. You obviously have a point," Dr. Davalloo said.

"Yes, I do," Liz said. "I had an interesting conversation with one of our meteorologists, and he explained to me that in the old days before the advent of gas detectors, miners used canaries to alert them when there was gas in the mines. I was thinking that must have happened here or something similar. The animals must have sensed or smelled the virus and fled the area. It's the fight or flight response all living creatures have when we detect danger."

"Interesting. Very interesting," Nikita said.

"It's getting close to dinner. Let's eat and discuss this some more. Plus, I need the time to check something out."

When everyone left the hub, Dr. Davaloo sat at the control console, looking around as if she were searching for something.

After getting their tray, Scotty, Professor Fehlinger, Ivanov, and Liz sat together. "Honey, you may be on to something," Professor Fehlinger said.

"I think I am Dad. Human beings are creatures of habit, and so are animals. There should be animals here, or at the very least signs of animal activity or passing. There's nothing. Something spooked them."

"What do you think spooked them?" asked Scotty.

"I honestly don't know," answered Liz.

"I thought about what you said about all living species having a fight or flight response. Is it possible that a predatory animal chased the other animals away?" asked Scotty.

"It's possible, but highly unlikely," Ivanov said. "There were polar bears here, and other than men, they have no other rivals. Besides that, all animals mark their territory, and Ms. Fehlinger said when she was walking around the perimeter she didn't see any evidence of animal markings or droppings."

"Excuse me. May I sit down?" Nikita said.

"Of course," Professor Fehlinger said. There weren't any more seats available at the table. Nikita took a chair from another table and sat next to her countryman, Ivanov.

"That was a very interesting theory you presented, Ms. Fehlinger. Can you tell me more?" Nikita said.

"I'm afraid not," Liz said. "We were just discussing the issue when you joined us."

Meals had a thirty-minute allotment time that was casually observed. Progression, regression, experimentation, and observation had their own time tables, and oftentimes scientists ate at all hours of the day and night.

When they returned to the hub, Dr. Davalloo was already there. Sitting next to her was Skip. Liz had met him earlier in the afternoon. Skip nodded at Liz as Dr. Davalloo introduced him. After the introductions were made, Dr. Davalloo explained his presence at the table. "Skip Wilder is our leading meteorologist. I asked him here because of the observations made by Ms. Fehlinger. When I first arrived here to investigate what happened, I found these papers." Dr. Davalloo stopped long enough to hold up a stack of papers before continuing, "They were neatly stacked on the console, which I thought was a little odd. There were weather reports. I believe whoever placed them on the console wanted to be found. I had the reports looked at earlier in an abstract, mundane sort of way. However, I found what Ms. Fehlinger said so compelling that I asked Mr. Wilder to take a look at the reports again for any anomalies. What he found could be a breakthrough, Mr. Wilder."

"Take a look at this," Skip said, standing up and leaning over the table as he spread out a map. "Here is where we're at," he said, pointing with his finger. "On our right is the Filchner Ronnie ice shelf, and on our left is the Ross ice shelf. Both of these shelves are fed by huge glaciers. Because of global warming, these glaciers have been melting at an alarming rate. When there is no wind, the water vapor from the glaciers dissipates into the Atlantic Ocean. However, when there is a strong wind, the water vapors from the glaciers are carried from north to northwest. I looked at the weather reports given to me by Dr. Davalloo, and then I went back several years looking at other weather reports and comparing them against the pattern of the virus appearing and reappearing. What I found was a perfect match."

"I found what Ms. Fehlinger said very interesting too, but the hypothesis is scientifically impossible," Sergi said. He was a Russian pathologist, and the first Russian Dr. Davalloo met after their initial contact on the WRC short-wave radio. "Granted, viruses are oftentimes airborne. That's just one of the many ways to contract a virus. However, viruses are not living organisms. They can only metabolize and reproduce when they are within a living cell. All viruses have to have a host. A living host. All this talk about glaciers and ice shelves is pointless. They're not living organisms."

"I can't speak to that. I'm not a scientist," Skip said. "What I can tell you is this: I looked at the weather reports detailing the conditions when the ship Martha was found, and she was sailing right into a northwestern wind current."

"There's a way we can prove or disprove the validity of my hypothesis," Liz said.

"How?" asked Scotty.

"Animals," answered Liz. "Animals have a heightened sense of smell than we do. They would have smelled the virus long before the infection started and moved to a place of safety. Mr. Wilder, how far would you say the last wind current blew before dissipating?"

The hub grew quiet as Skip looked at the weather reports while jotting down calculations on a notepad. After a few minutes, he said, "Approximately sixty miles due south."

"Okay," Liz said. "We need to find out where the animals went. If we find them somewhere in the radius of sixty miles in places they should not be, then we can surmise that for whatever reasons this virus is airborne," she finished, looking around at the faces sitting at the table.

"That sounds like a good plan of action. I suggest we eat, then we can research what animals are native to this area, get some sleep, and be ready to venture out first thing in the morning," Dr. Davalloo said.

"In the meantime," Scotty said, "I'll get in contact with the CDC and the White House to give them an update on what we're doing. After that disastrous news conference, I'm sure the president will appreciate any new developments."

Chapter 10

The next morning was a clear, freezing, cold, sky blue day. There was no wind. After eating breakfast and drinking hot cups of coffee, they left the warmth of the WRC. Scotty, Professor Fehlinger, Liz, Dr. Davalloo, and Skip crowded into one of the snowmobiles, while Sergi, Legkov, Vladimir, Nikita, and Lydia crowded into another. They drove in opposite directions. Skip drove north. Sergi drove south.

There were no tracks of any kind to follow. Skip drove, following the wind currents he had mapped out. The vast wilderness of Antarctica was uncharted. The snowmobiles had to proceed slowly to avoid spin-outs from the ice underneath the snow and from sinking where the ice was too thin to hold the weight of the snowmobiles. As they moved slowly along, Skip charted their course so they would not get lost returning to the WRC.

"We are past the sixty-mile radius and still have not seen any animals," Skip said, voicing what everyone was thinking.

"Just a few more miles," Scotty said, not wanting to give up so easily.

They drove a few more miles, and slowly, they started seeing all sorts of animals in a hodgepodge. The scene was breath-taking. It was like stumbling upon the Garden of Eden, except there were no trees. Just snow and ice and ponds of water. Skip stopped the snowmobile. They all looked around. No one said anything for several seconds. "Is this natural?" asked Dr. Davalloo.

"What do you mean?" Liz said.

"For all of these animals to be here like this, different species, cohabitating."

"That's a question we will have to find an answer to later," Scotty said. "The more important observation is that Ms. Fehlinger's hypothesis seems to be correct, which means the virus is airborne. We have to find its source of transmission."

"I may be able to help with that," Skip said. "Once we get back to the WRC, along with my team, using mathematical equations, we can go over the weather reports and calculate the outward swirls back to their origins."

"How precise are these calculations?" asked Scotty.

"Weather reporting is not an exact science because of all the variables involved that could cause a sudden shift. Nonetheless, we can get pretty close."

"Okay," Scotty said. "Let's take some pictures and get back to the WRC."

The snowmobiles had a four-sided thick plexiglass enclosure that allowed for a panoramic view of the outside. Some of the animals came up to the snowmobile and sniffed around, but most of the animals ignored them.

When they finally returned to the WRC, the sun had started to descend, and their bladders were screaming for release. After a long trip to the bathroom, the two snowmobile research teams compared notes, pictures, and the significance of the discovery of the animals. The Russians had also found a Garden of Eden to the south.

Skip, along with his meteorologist team, poured over barometric pressure weather reports, wind reports, wind currents and their speed, temperature, and mean reports. On the third day, Skip sat down with Dr. Davalloo, Sergi, and the team of other scientists and said, "I looked at all the weather reports starting in 1957, when the WRC was first established, until now. What I can tell you with confidence is that each year, based on those reports, the earth is progressively getting warmer, and what I can also tell you with confidence based on these reports and my calculations of the wind currents is that I believe the source of the virus transmissions are the glaciers."

The hub was quiet as everyone sitting at the table digested the information and what it meant. Scotty was the first to speak.

"Assuming what we just heard is true, how is it possible for a virus to survive in unanimated matter?"

There was more silence before Professor Fehlinger said, "There is only one way. An anaerobe"

"Of course!" Hal said. "I completely forgot about the anaerobe."

"An anaerobe? Please explain. What is this anaerobe?" Sergi said.

"An anaerobe," Vladimir said, "is a little-known living one-cell organism that can survive in an environment free of oxygen or air."

"I don't understand," Liz said. "I thought viruses could only grow and multiply inside a living host."

"Technically, that's correct. However, in the world of viruses, there are anomalies," Professor Fehlinger explained.

"This could very well be one. I think we should take samples from a glacier and test for any presence of an anaerobe now that we have at least an idea of what we are looking for," Dr. Davallo said.

"I agree," Professor Fehlinger said. "If we can get a sample from a glacier, we should be able to put the sample under a microscope and see what's there. There may not be any anaerobes. I just offered that as an explanation of how something could survive in an oxygen-free environment. The bigger problem now that we've decided on a course of action is a logistic one."

"How so?" asked Scotty.

"Remember the Titanic," Professor Fehlinger answered.

"Sure. It was the biggest maritime disaster in America's history."

"Right, and do you remember why the ship was sunk?"

"It was hit by an iceberg," Scotty answered, a little annoyed by the question he felt any junior high school student would know.

"That's right," Professor Fehlinger said. "The top half of the iceberg was visible, but the bottom half wasn't. It was underwater, and that was the part where the Titanic struck. We'll be running the same risk if we try to negotiate a ship or boat close enough to get to a glacier to take samples. They usually have icebergs floating around them. The water is too cold for divers. They would die from hypothermia before they could reach a glacier."

"What about a helicopter?" asked Scotty. "We could fly a helicopter over a glacier and then lower someone down to take the samples."

"That's an excellent idea," Professor Fehlinger said, nodding his head. "We can use some of the equipment I've seen here to analyze the samples."

"How soon can we get a helicopter?" asked Dr. Davalloo.

"The CDC is too far away. We'll need the coast guard to help us. I can radio them and tell them what we need. I'm sure they can handle the technical stuff. So, I'm thinking..." Scotty paused, rocking his head from side to side as if he was counting before continuing, "Sometime tomorrow morning. Maybe in the afternoon at the latest."

"While you're taking care of that, the rest of us will start preparing the lab for samples," Dr. Davalloo said.

It was late in the morning when the coast guard helicopter took off from its base in Anchorage. The base was operating at minimal capacity. Most of the guards, men and women, were stricken with the virus, and getting a helicopter in the air had taken a call from Scotty to the Secretary of Defense, Thomas Reed, to get one in the air. Instead of a five-member crew, there were only three on board. Lieutenant Baylord Diggs was the pilot. There was no copilot. Sergeant Troy Milken was the crane operator, and Private Calvin Smith was the point man for the operation.

Private Smith was wearing a hazmat suit and skin-tight latex gloves. Attached to the front of the hazmat suit was a vest with several horizontal-shaped clear plastic tubes. Also attached to the belt was a hand-held battery-powered drill.

The helicopter flew over two huge glaciers and hovered.

"Ready?" Sergeant Milken shouted over the helicopter rotary blades.

Private Smith gave the thumbs-up sign.

"Remember. As I start lowering you at any time, you want me to stop yanking hard once on the cable. When you're ready to come back up, yank two times."

"Got it!" Private Smith shouted. He held onto the cable as Sergeant Milken began lowering Private Smith down to one of the glaciers. Private Smith dug the toe of his spiked snow boots into the snow and icy surface of the glacier. When he was sure of his footing, he yanked once on the cable and disengaged.

Private Smith walked along the snowy and icy surface of the glacier, listening to his footsteps crunch in the eerie silence. After a few feet, Private Smith knelt and drilled a hole in the surface. He carefully inserted the tube into the hole and pulled it back out once it was filled with snow and ice. Private Smith placed a rubber cap on the tube, sealing it. Private Smith repeated the process two more times, each time walking further out. When he was finished, Private Smith retraced his steps to the dangling cable, and once he was secured in the harness, he yanked on the cable two times.

The helicopter slowly pulled Private Smith up enough to clear the towering glacier and flew to the next glacier. Sergeant Milken lowered Private Smith along the side of the glacier, where he dug his spike boots into the side of the glacier. While the cable and spiked snow boots held Private Smith securely in place, he drilled a hole into the side of the glacier. After drilling the hole, Private Smith inserted the tube into the hole until it was filled with snow and ice and placed the rubber cap on the tube, sealing it. Private Smith swung around the glacier like a lumberjack, taking samples of the glacier from different areas. When the tubes were filled, he yanked twice on the cable. Sergeant Milken slowly pulled him up. Once Private Smith was back on board, Lieutenant Diggs flew to the WRC.

Dr. Davalloo and Scotty were there to greet Sergeant Milken and Private Smith. Lieutenant Diggs stayed in the helicopter. It would be a quick landing and takeoff. After Private Smith deposited the glacier samples with Dr. Davalloo, Scotty thanked them and the coast guard for their assistance before they boarded the helicopter.

Dr. Davalloo carried the samples into the lab, where the scientists were waiting. Although Liz's observations were the impetus for the collection of the samples from the glaciers, she was not there. She was not a scientist. As a precaution, the scientists put on hazmat suits before opening the tubes.

The lab was equipped with three microscopic stations with pinchers to handle samples without human contact. They would have to share. There were twelve tubes.

"Ready?" asked Dr. Davalloo.

They all nodded their heads, not knowing what to expect.

Dr. Davalloo passed out the tubes evenly. Six tubes were given to the American scientists' team, and six were given to the Russian scientists' team.

Dr. Davalloo emptied one of the tubes onto a glass side and resealed it with the rubber cap. Dr. Davalloo peered into the microscope as she adjusted the dials for maximum magnification. The pathogen came into focus. It looked like a black, lumpy ball with spikes sticking out of it from every angle and tiny hooks on the end. Dr. Davalloo stepped away from the microscope and said to Professor Fehlinger, "Take a look."

Professor Fehlinger peered into the microscope. After a few seconds that seemed like minutes with Professor Fehlinger making his own adjustments, he stepped away from the microscope and said, "Ugly little buggers. I've never seen a pathogen like that before. It looks...angry. Like its exploding."

"Neither have I," Dr. Davalloo said, as Scotty peered into the microscope.

"Let's run it through the spectra," Dr. Davalloo said.

"Okay," Professor Fehlinger said.

The spectra was short for spectrometer. It was an instrument originally designed for measuring the spectrums in color by a beam of white light dispersed so that its parts are arranged in order of their wave lengths. It was discovered in 1947 that a beam of radiation dispersed on a specimen would create electromagnetic markers that could be used to map a specific pathogen. Each pathogen has its own character trait that enables scientists to develop a specific vaccine for that pathogen.

The other scientists, including their Russian counterparts, were studying the pathogen, trying to decipher its genetic structure. What was left of the remaining day and the next two weeks were spent conducting a battery of tests. All of the tests came back negative or conclusive. The scientists found it frustrating watching the virus continue to spread and the mortality rate increase and not being able to do anything about it after discovering the causative pathogen.

The WRC cafeteria had become the social gathering area for the scientists. It was another long night conducting tests and peering into microscopes when a few of the scientists sat scattered around in the cafeteria voicing their frustrations.

Dr. Davalloo, Professor Fehlinger, Scotty, and Liz sat together. Since their arrival, Professor Fehlinger and Dr. Davalloo have worked more closely together than the other scientists. Their affinity didn't go unnoticed. Liz was happy to see her father showing interest in a woman. She knew it had been a long time since her father had shown any romantic interest in another woman since her mother had died.

Dr. Fehlinger was also making his own observations about how close his daughter and Scotty were becoming. They made a good couple, and times were changing about interracial relationships.

"A penny for your thoughts," Scotty said.

"I'm sorry. My mind was elsewhere," Liz said.

"I could tell that," Scotty said. "What I was saying is how frustrating this situation is. We know what the pathogen is that is causing the virus. We know that it's airborne, and its source. After that, we seem to have hit a brick wall. None of the vaccines we have used in the past against other airborne pathogens neutralizes this one."

"Add to that," Dr. Davalloo said. "We have not been able to date this pathogen."

"Why not?" asked Liz.

"That's a complicated answer. Without getting bogged down in details, we have a machine called a spectra that bombards a pathogen with radiation, and when that happens, the radiation that bounces off the pathogen gives off an electromagnetic wave that is distinct to that specific pathogen. We then compare that pathogen against all known or similar pathogens in that family to develop a vaccine, and so far, we have not even come close to finding a distant cousin. It's as if this pathogen popped up from nowhere."

The table fell silent as everyone felt the weight of their own frustration.

"What about the glaciers?" asked Liz.

"What about them?" Dr. Davalloo said.

"Is there any way to date a glacier?"

"I don't know," Dr. Davalloo said, her forehead wrinkling in thought. "I know someone who might know. Excuse me," she said, getting up from the table.

Dr. Davalloo went to the hub of the WRC, where she knew Skip would be preparing weather reports.

"You're up late, or shall I say early?" Skip said as Dr. Davalloo stood next to him.

"We've all been firing on all cylinders since we found the pathogen."

"I heard about that. How's it going?"

"Small steps. Actually, I was hoping to find you here. I'm optimistic you can help me."

"Whatever I can do," Skip said, turning toward Dr. Davalloo and giving her his full undivided attention. It was the first time any scientist had ever sought out his help.

"Is there any way to determine the age of a glacier?"

"Not exactly, but we can get really close. Why do you ask?"

"We haven't been successful in dating the pathogen, so I was thinking that since the pathogen was found inside the glacier, if we can date the glacier's age, it'll give us an approximate date of the pathogen."

"Clever," Skip said. "We'll need a petrologist."

"A what?"

"A petrologist. It's a person who studies rocks. Glaciers are made up of rock, ice, and snow that solidify over time. I'm speculating that the same techniques used for dating rocks can be used for dating any kind of solid environmental surface."

"Thank you," Dr. Davalloo said. She left the hub and found Professor Fehlinger, Scotty, and Liz still sitting at the table. Dr. Davalloo reiterated what Skip had explained to her. When she finished, Scotty said while getting up from the table, "Good work. I'll make some calls."

Chapter 11

The steady, rapid knocking at Malone's front door awakened June first. She nudged her husband, Kent, awake.

"Honey, there's someone knocking at the front door."

"What? At this time in the morning," he said, reaching for the clock on the nightstand next to their bed. It was three o'clock. They both heard the knocking again.

"Somebody better be dead or dying," Kent said, annoyed at having to get out of bed.

Kent flicked on the house lights as he went down the stairs. The knocking came again. More insistent this time. Kent peered through the door peep hole. He saw three men dressed in military uniforms. One had the insignia of a lieutenant. He was doing the knocking.

Kent opened the door.

The lieutenant and the other men showed their identification. As Kent was looking over their identification, the lieutenant said, "I'm sorry to disturb you at this hour, but we have a matter of national importance and we need to speak to your wife. May we come in?"

"Sure," Kent said as he stepped aside, allowing the men to enter. Once they were inside the house, Kent returned their identification. The lieutenant's name was Herman Louse, the sergeant's name was Oreader Cole, and the private's name was Martin Styles.

"What's so urgent you have to speak to my wife about at three o'clock in the morning?" Kent said to the lieutenant.

"Mr. Malone, believe me, I understand your concern. However, this is a matter for your wife."

"I'm June Malone. Whatever you have to say to me, you can say in front of my husband," June said, after hearing voices and coming down stairs. She stood next to Kent with her arms folded across her chest.

"Mrs. Malone, I'm here on a matter of national importance. As the foremost leading petrologist in the country, your expertise is needed. I'm here to fly you to a remote research center stationed in Antarctica."

"The Antarctica!"

"Yes, ma'am," Lieutenant Louse said. "The sooner we get started, the better. It's my understanding that your assistance is urgently needed."

"I can't just…"

"Mrs. Malone, please excuse me for interrupting you. We really don't have a whole lot of time to debate this. I was instructed to tell you that the work you are being requested to assist with involves stopping the spread of this pandemic."

June looked at Kent. "I can take care of things here," he said.

"Okay," June said. "I'll need time to pack some warm clothes. How long will I be needed?"

"I don't know, ma'am, but having done these types of missions before, my advice is to pack enough clothes to last at least a week," Lieutenant Louse said.

June went upstairs to pack, followed by Kent. She packed two suit cases of winter clothes. When she was finished, Kent carried the suit cases down the stairs and set them by the door. He turned to June, and they embraced tightly. "I'll call you as soon as I can. Try not to worry too much about me," she said.

"I'll make a deal with you. You take care of yourself, and I'll take care of myself," Kent said.

"Deal," June said, smiling. Letting go.

Private Styles carried the suit cases out to a four-door dull green-colored *Ford* army car and put them in the trunk.

Lieutenant Louse got behind the steering wheel. Next to him sat Sgt. Oreader Cole. Sergeant Cole picked up the car radio and said, "We have the package, and we're on our way to Vandenburg's air base." Lieutenant Louse started the car and pulled out into the street.

The Malones lived in Lompoc, California. Vandenburg's Air Force Base was in Santa Monica. Thirty miles away. Lieutenant Louse sped along the streets, ignoring traffic lights. At three thirty in the morning, there were no other vehicles to contend with. They reached Vandenburg's air base in twenty minutes. A plane was waiting for them on the tarmac. They left the car and boarded the plane. On the plane, June learned Private Styles was with the army's engineer and was assigned to assist her in any way she needed his assistance. June also learned on the plane that there was an electron diffraction machine and a generator to power it.

The plane landed at the coast guard base in Anchorage, where they boarded a helicopter for the final flight to the research station. When the helicopter touched down, Scotty and Dr. Davalloo drove out on a snowmobile to meet them. Lieutenant Louse, Private Styles, and June departed the helicopter. Lieutenant Louse introduced himself, Private Styles, and June Malone. Dr. Davalloo introduced herself, and then Scotty. After the introductions were over, Lieutenant Louse, Private Styles, June, and Dr. Davalloo transferred the electron diffraction machine and generator from the plane onto the snowmobile. When the task was completed, Lieutenant Louse said, "I have to get back to base. Private Styles will stay on as Mrs. Malone's assistant in any capacity as needed. Any questions?"

"I'm sure we can manage things from here. I and we really appreciate your assistance on such short notice," Scotty said.

"Anything I can do to help stop this scourge," Lieutenant Louse said as he returned to the helicopter.

The electron diffraction and generator made it a tight fit for everyone to fit in the snowmobile. The short drive to the WRC was uncomfortable, as the sharp corners of the electron diffraction and generator could be felt poking through the thick winter clothing as they rode over packed snow and ice. When they arrived, all the scientists were in the hub. Dr. Davalloo made all the introductions. After the introductions were over, Dr. Davalloo showed June where she would be sleeping. After depositing her luggage, Dr. Davalloo showed June the lab. Before entering, they put on hazmat suits. "This is why we need your help. Take a look," Dr. Davalloo said.

June peered into the microscope. She had no idea what she was seeing. "What am I looking at?" she asked.

"It's called an anaerobe," Professor Fehlinger said, entering the lab. He explained it's unique characteristics. When Professor Fehlinger finished, June said, "I'm sorry. I think there has been a mistake. I'm not a biologist. My field of study is rocks. I'm a petrologist. I don't understand how I can be of any assistance here."

"What we are hoping for," Dr. Davalloo said, "is for you to give us your best approximate date of the glacier the anaerobes were found in. Is that something you can do?"

June didn't say anything at first. Dr. Davalloo had presented her with a question she had never thought about before. She thought about the properties of a glacier. Although a glacier is mostly formed by time, pressure, snow, and ice, there would be rocks trapped in the layers, and she could date those rocks, she thought. "I think I can," June said finally. "I've never done anything like this before. I'll give it my best shot."

"Thank you," Dr. Davalloo said.

There was no room in the lab or hub to set up the electron diffraction generator. Private Styles, with help from some of the Russian scientists who were very much interested in America's technology, set up the electron diffraction and generator in the utility shed behind the WRC. Afterwards, June and Private Styles settled into their sleeping quarters and went to sleep. Tomorrow would be a long day.

When June and Private Styles awoke, they ate breakfast and immediately went to work. The coast guard had delivered blocks of snow and ice taken from where the glacier samples were taken and left them outside in the freezing cold next to the utility shed.

Private Styles, using a pick and hammer, chipped off pieces of the blocks of snow and ice and carried them into the shed, where they were placed inside the electron diffraction steel box. The electron diffraction directed a beam of gamma rays on the snow and ice, illuminating its layers and exposing rock sentiments created by accumulations of deposited materials over the years. The deposits had slowly cemented together over long periods of time, capturing clues to the earth's geological clock. In an undisturbed sequence of snow and ice layers, the lower layers of material found are older than the materials found in the upper layers.

June ignored the materials found in the upper layers since any materials found there would have been recent and the organism identifiable. Instead, she focused the gamma rays on the lower layers.

The work was difficult and cumbersome. They had to wear hazmat suits while working, and the gamma rays melted the snow and ice. Private Styles had to constantly supply June with snow and ice from the blocks taken from the glacier. June went through several blocks each time, picking up on the layer where she had left off before the snow and ice melted. What June found in those lower layers were rock sentiments, flowers, fossilized vertebrates of small animals, and a variety of plant life on the ground level of the snow and ice before the glaciers were formed. The materials found from June's trained eye peering through the microscope were from the holocene period, extending from twenty thousand to eleven thousand years ago.

After approximating the date of the glaciers, Scotty radioed Lieutenant Louse to fly June back to her home in California, away from the freezing cold, where her husband was anxiously awaiting her safe return.

They were all sitting at the huge round table in the hub. The American scientists, the Russian scientists, Dr. Davalloo, Scotty, Professor Fehlinger, and Liz. Twenty to eleven thousand years ago. Professor Fehlinger had heard those specific numbers before, but where? He was plagued by the thought.

"We know of pathogens far older than twenty or eleven thousand years old. The ancient world knew about skin lesions, small pox, arthritis, and other diseases, which means this pathogen is unique to those diseases, which further complicates finding a vaccine. Professor Fehlinger. Are you with us?" Dr. Davalloo said, noticing the vacant look on his face.

"Yes," Professor Fehlinger lied, wanting to keep his thoughts private until he could understand why those numbers gnawed at him. "You were saying how the ancient world knew of diseases that were far older than the pathogen we are dealing with, which illustrates its uniqueness. Does that about sum up what you were saying?"

"Exactly," Dr. Davalloo said. "Liz, your idea got us this far. Any more?"

"I wish I did," Liz said.

The meeting ended in frustration, which everyone was becoming accustomed to. As soon as the scientists left the hub, Skip and his team of meteorologists moved in to begin plotting the course of the weather.

It was early—three o'clock in the morning—when Professor Fehlinger suddenly sat up and looked around the dimly lit dormitory. Not all of the cots were occupied. Some of the scientists were still in the lab. Some were in the cafeteria, while others were involved in a secret romantic rendezvous. There were no rules against dating. Scotty's cot was next to Professor Fehlinger. He was lying on his side, asleep. Professor Fehlinger remembered why those numbers of twenty and eleven thousand were so significant to him. Professor Fehlinger felt an urge to awaken Scotty but didn't. Instead, he threw back the thick blanket and got dressed.

Professor Fehlinger went to the hub. He saw Skip and his team of meteorologists pouring over weather reports and graph charts. They were having a serious conversation about something. He could tell by how animated they were and how seemingly oblivious they were to his presence. Professor Fehlinger clicked on the WRC intercom system. "Dr. Davalloo. Please report to the cafeteria. ASAP." Professor Fehlinger repeated the request two more times to stress its urgency before clicking off.

Professor Fehlinger was on his third cup of coffee when Dr. Davalloo entered the cafeteria, looking around. She saw Professor Fehlinger and went straight to his table. "I thought I was dreaming when I heard you on the intercom," she said, sitting down.

"I'm sorry, but I thought this was something you would want to know immediately. I think, and this is a long shot, but it's possible I may have found what was once a living host for the pathogen."

"That's great!" Dr. Davalloo said excitedly. "We can..."

"Whoa," Professor Fehlinger said, holding up his hands and interrupting her. "I told you this was a long shot. A very long shot. Did you know I was married once?"

"I made that assumption since you have a daughter," Dr. Davalloo said.

"It was a long time ago. She was Liz's mother. Her name was Elizabeth. Liz is named after her. She was an anthropologist. Before she died, she was working on the hypothesis that there were people who lived in Antarctica thousands of years ago before it became covered in snow and ice. She found the fossil remains of a man buried in ice dating back to twenty to eleven thousand years ago. As human beings, we all carry bacteria, viruses, germs, etc. from the environment in which we are born. Our bodies become asymptomatic to these maladies. Otherwise, we would constantly die off as a species, as some suspect befell the Neanderthals. It's my guess the fossil remains my wife found could very well be the host to this pathogen, and if so, his body would have developed antibodies to fight off the pathogen and stop the virus from reproducing itself."

"If we can extract the antibodies from his DNA tissue, we can develop a vaccine," Dr. Davalloo said excitedly. It was the first glimmer of hope she had since arriving at the WRC.

"That's the long shot," Professor Fehlinger said. "Actually, the whole thing is a long shot." Davalloo was silent for a few seconds and then asked, "That's what you were so preoccupied about yesterday at the meeting?"

"Yes, but at the time I didn't understand why. It wasn't a good time in my life. After the meeting, I went to sleep, but something kept nagging at me until I remembered why," Professor Fehlinger answered.

"Sterling," Dr. Davalloo said as she reached out to hold his hands while calling him by his first name. "I don't want to seem insensitive, but we need to move on this as quickly as we can if there is the slightest chance of developing a vaccine. Where is this fossil now?"

"The sitting president at that time decided global warming was a hoax, so his administration defunded any programs having to do with climate change or its impact on people, essentially destroying my wife's lifelong work. The fossil she discovered was ultimately shipped to the Smithsonian Institute in Washington, D.C."

"Okay. We have to get Scotty on board with this. He'll know who to contact to get us access to the fossil. We may need your daughter's expertise on this as well."

"I'll bring Scotty up to speed while you brief my daughter."

Chapter 12

John Paul was the curator of artifacts at the Smithsonian Institute in Washington, D.C. It was seven thirty in the morning eastern standard time when he got the call from Scotty. John Paul listened intently and then said before hanging up, "I'll make the necessary preparations, and we'll be expecting your arrival."

The Smithsonian Institute is a large building. John Paul's office was in the basement, along with the Institute's most important and rare massive collection of artifacts the world over. John Paul had a staff of twenty-five curators under his supervision who also worked in the basement before the virus. Now there were only two curators under his supervision.

After Scotty's call, John Paul accessed his computer inventory until he found the fossil dubbed the Ice Man. John Paul read the attached file about the institute's procurement of the fossil. When John Paul finished, he picked up his walkie-talkie on his desk and said, "Vargas, I need to see you in my office, please."

"On my way," a female voice responded.

Ten minutes later, there was a knock at John Paul's office door. "Come in," he said.

K. Vargas entered the office. Everyone except John Paul called her K. He thought, as her supervisor, it was inappropriate. "Sit down. Please," John Paul said.

After she had sat down in a chair across from John Paul's desk, he said, "I need you to prepare the ice man's fossil for a DNA extraction tomorrow morning. I'm not sure of the exact time, so just be ready for any time before noon. This is a priority. Understand?"

"Understood sir. Anything else?"

"No. That will be all. If you have any problems, radio me immediately."

The Smithsonian Institute exhibits were arranged in sequential numerical order, with an alphabet assigned in front of the number indicating where the exhibits were located. The letter B meant the exhibit was located in the basement, and the number 861 was the row or area in which the exhibit was stored. B 861 was a refrigerated storage unit for exhibits requiring placement in a control environment.

When K. Vargas left John Paul's office, she radioed for the other curator to assist her. Before entering what they called "the box," where fossils and other rare ancient world artifacts were stored inside a regulated temperature control environment to combat deterioration caused by heat, they sterilized themselves and put on hazmat suits to prevent cross-contamination. After checking to verify Ice Man's catalog alphabet and number were correct so they wouldn't have to scramble to locate the exhibit at the last minute, they began preparing the examination room where the DNA extraction would take place. They scrubbed and sterilized the room from top to bottom and lowered the room's temperature to acclimate it to the Ice Man's temperature-controlled environment. When they were finished, K. Vargas locked the door closed and hung a *do not enter* tag without *proper authorization* on the door knob.

Dr. Davalloo, Scotty, Professor Fehlinger, and Liz were picked up by a coast guard helicopter and flown to its base in Anchorage, where a military plane was waiting for them to fly them to Reagan's Air Port in Arlington, Virginia. At the airport, an army car was waiting for them for the short drive across the 14th Street bridge into Washington, D.C. The Smithsonian Institute was located on Maryland Avenue South West.

When they arrived, John Paul was waiting for them in the lobby. Dr. Davalloo, Scotty, Professor Fehlinger, and Liz had all visited the Smithsonian before, but to see it empty because of the virus was an eerie feeling surrounded by all the exhibits. It felt like they were being watched and judged by them.

Scotty made all the introductions after John Paul introduced himself. Once the introductions were over, John Paul said, "I know time is a precious commodity. Please follow me. Everything has been arranged."

"I want to thank you for accommodating us on such short notice," Scotty said as they followed John Paul through a maze of corridors until they stopped at an elevator. They rode the elevator to the basement. They followed John Paul to a scrub room, where they all scrubbed and donned hazmat suits.

They entered the examination room across from the scrub room. The fossil of Ice man's remains were laying on a table. Next to the table was a covered tray. At first, they all just stood there looking at the Ice man. He seemed to be at peace frozen in time. Professor Fehlinger had never seen his wife's discovery before. He was waiting for the big event. His wife had aspirations of finishing her research then presenting the Ice man to the public at a large exhibition, but never got the chance because her funding was cut short. Professor Fehlinger felt a strange connection to the Ice man. Professor Fehlinger found it ironic that in some strange way his wife discovery of the Ice man which now he understood contributed to her death now could hold the key to saving all their lives. Professor Fehlinger looked at his daughter. He wondered what she was thinking.

Liz was too young to remember or understand the significance of her mother's discovery and research. Liz had wanted to see the Ice Man, but access to him was denied to prevent cross-contamination.

Liz moved slowly at first, walking around the table. As Liz walked around the table, Dr. Davalloo, Scotty, and Professor Fehlinger all began walking around as if they were staking out their positions. Liz stood on the left side of the table at the top, near the Ice Man's head. Next to her was Scotty. Across from Liz was Dr. Davalloo, and next to her was Professor Fehlinger. John Paul had stood away from the table as an observer.

All living things, from human beings to bacteria, have nucleic acids called DNA and RNA, which are hereditary materials that direct the operation of cells. An alteration in an organism's genetic code is called a mutation. This mutation changes the order of amino acids in a protein and effects the biochemical properties of the protein, changing it into a pathogen. However, this change does not kill the host because the mutation takes thousands of years to occur, allowing the body's immune system to develop T cells to specifically fight off the pathogen altogether or force the pathogen to lay dormant until a lifestyle change or something in the environment weakens the immune system and allows the pathogen to resurface and spread beyond the host.

The basic principle of Darwin's theory of evolution is that all living creatures adapt to their environment by inheriting genetics from their ancestors to survive under the conditions of that environment. When this doesn't take place, those species come to a genetic end and die out.

When the Ice Man lived, Antarctica was not a complete snow and ice land with large glaciers and ice shelves. The landscape was lush with green grass, trees, and all sorts of wild animals. Ice Man's ancestors genetic code would have had to mutate to allow him to survive once Antarctica's weather conditions became harsh. The Ice Man was found in sedimentary rocks, which were created by an accumulation of deposited materials. The deposits slowly cemented together over long periods of time, capturing fossils, flowers, and other clues to the earth's geological history. The Ice Man had lain undisturbed while snow, ice, glaciers, and ice shelves formed all around and over him layers upon layers.

Through a process called amino acid racemization that uses amino acids, which are the building blocks of protein, scientists can determine that two forms of L and D amino acids in living things are of the L form. At death, the L form racemizes into the D form with strands of T cells, which are called trace fossils and act as antigens to fight off pathogens of their time.

Dr. Davalloo removed the small square-shaped cloth covering the tray next to the table, revealing the instruments needed for the DNA extraction. Dr. Davalloo picked up a needle and paused as if she were making a silent prayer. She understood how much was at stake in what she was about to do. She took a deep breath, exhaled slowly, steadied her hand, and slowly inserted the needle into the fossil's nerve sac.

Chapter 13

Geography, in its broadest definition, is the systematic study of the earth's physical makeup and how those elements affect the way humans organize their communities and live productive lives.

Geographers examine human populations and population distribution by region, urban and rural segmentation, political systems, and economic structures.

Meteorology is the science that deals with the atmosphere, its phenomena, and weather forecasting.

Climate is defined as the characteristic condition of the atmosphere near the earth's surface in a particular place or region. It includes patterns of temperature, wind, precipitation, cloud cover, and other variables. Earth's climate is ultimately dependent on solar radiation, its distance from the sun, the orientation of the earth's axis of spin away from or toward the sun, and other factors. Climate varies from region to region depending on latitude, the topography of the land, ocean water currents and temperature, and atmospheric pressure and wind. It can be predictable in terms of the average and variation of temperature and other variables, but is subject to change due to the long-term cycle of ice ages and human activities.

Skip was a climatologist. He understood the wind systems that were determined by variations in atmospheric pressure. The effects of global warming were creating a low atmospheric pressure in the Antarctica region known as the "roaring forties" and "furious fifties" that blows from east to west. The barometer pressure used for determining the weight of a column of air was gradually falling, indicating the building of a strong wind current.

Skip was worried. He was very worried as he looked at the data. Everyone at the WRC was aware of the pathogen found in the glacier and what it meant. Dr. Davalloo and Scotty were in Washington, D.C., on business.

Skip radioed the CDC about his concerns.

Bradley Cosgrove was the assistant director at the CDC. It was early in the afternoon when he received Skip's radio transmission. Bradley explained that Scotty was preoccupied with a very important project and would be unavailable for some time. Bradley offered his assistance. Skip couldn't wait until Scotty became available. It may be too late by then. Skip explained the nature of his call. Bradley listened intently, understanding the severity of Skip's observations about the strong wind currents that would be sweeping down Antarctica and bringing the virus with them.

"Have you talked to anyone else about this?" asked Bradley.

"No, I haven't. My intention was to alert Mr. Scott before you told me he was unavailable."

"Does anyone else know about this?"

"Only bits and pieces."

"I don't understand what that means."

"I have a meteorology team where each person performs a specific function. Once that function is completed, it goes into a report, and all the reports are turned in to me for final analysis," Skip explained. "No one knows the complete picture but me."

"Okay. I understand," Bradley said. "I need you to keep this information confidential."

There was a moment of silence before Skip said, "I won't put people's lives in danger."

"I understand, and I'm not asking you to do that. I'm asking you to keep the information confidential until I can reach Mr. Scott. I'm sure the people there have families, and like any other family, they talk among themselves, and if this information were to become public knowledge with people already dying and afraid..." Bradley paused, allowing the silence to make his point, and it did because Skip's next words were, "I understand. I won't hold the information for too long. There's only a small window to do something. I need to know the second you reach Mr. Scott and what the CDC intends to do."

"I will," Bradley assured Skip before hanging up.

Bradley was a security level six member of the secretive wing of the armed services known as Advance Forward, or AF, among its members. Very few people knew of its existence. They were part of every essential branch of government. Federal and state. Their mission mandate is to safeguard and protect the interests of the United States nationally and internationally in real time without having to be bogged down with legal technicalities like search warrants, arrest warrants, court-approved wire taps, or authorization from foreign governments to operate on their shores or waters. Because of these legal omissions, only a very select handful of people knew of their existence, and only a few of those had direct contact with Secretary of Defense Thomas Reed. Bradley Cosgrove was one of those few.

After talking to Skip Bradley, he immediately called the Secretary of Defense. Their conversations were always short and to the point, with no pretense at pleasantries, for a lot of reasons, but mostly because, in the transfer of highly sensitive information, time was a luxury they didn't have. The Secretary of Defense receiving a phone call from any of the select few who had his direct line meant the country was already facing a crisis that needed immediate action. When Bradley finished talking, the Secretary of Defense simply said, "Keep me posted on any changes," and hung up.

After securing the Ice Man's DNA and scrapings from what remained of skin tissue and bone, they were placed inside a small plastic red refrigeration box like the kind they use for organ transplants and taken under army escort to Reagan's Airport, where they were given access to a Cessna Jet and flown to the CDC in Atlanta.

When a bacteria or virus invades the body, they trigger a counterattack known in medical terms as an immune response. This response includes synthesizing antibody proteins that combine with antigen molecules on the surface of the invader, destroying the invader. Secondarily, the body forms long-lived memory cells. If, again, at a later time, the disease returns, the memory cells immediately kick into action, releasing antibodies that form the basis for vaccines for certain antigens. It's a long and laborious process involving time, sophisticated testing, and more testing before the vaccine can be developed and safely administered.

Dr. Davalloo and Professor Fehlinger spent most of their time almost working around the clock with a team of CDC scientists in an effort to fast-track a vaccine for the virus that was steadily killing people on their watch as if they were being taunted by the virus. Scotty used his position and influence to make sure the work they were doing was not bogged down by layers of bureaucracy. Scotty had left the mundane day-to-day operations of the CDC to his assistant, Bradley Cosgrove.

The Food and Drug Administration, more commonly known by its acronym FDA., was located on New Hampshire Avenue in Silver Spring, Maryland. On April 11, 1953, Congress created the Department of Health, Education, and Welfare. On September 27, 1979, Congress approved the creation of a separate Department of Education. The original Department of Health, Education, and Welfare was renamed the Department of Health and Human Services, which was an umbrella agency that administered a wide range of programs in the fields of health care and social services that affected nearly all Americans. Under its umbrella is the FDA, which assures the safety of food, drugs, cosmetics, biological products, and medical devices.

Scotty was on the phone talking to Edmond Baily, the Director of the FDA, about the bureaucracy involved with developing vaccines when he saw the red light on his phone panel start to blink. Scotty knew for Bradley to call him it was an emergency above his pay grade.

"Edmond, I have to call you back. I have an emergency coming through on another line."

Scotty clicked off and pushed a button, switching over to another line. "What's going on?" he said.

"Sorry to have to bother you, boss, but I received some very disturbing news from the research station in Antarctica that I think you should be aware of." Bradley told Scotty what Skip had told him.

After talking to Bradley, Scotty hung up. He sat there in his office, thinking. They were on the verge of possibly finding a vaccine for the virus, and now this. All the preliminary research markers indicated they were on the right track. What they desperately needed was more time. Until now, they didn't have. If Skip's warnings were true, the wind currents were picking up and sweeping down from Antarctica carrying the deadly virus, which would possibly decimate large populated areas of the United States eastern and northern borders. The virus had already weakened the United States economy. If Skip's predictions were true, it could plunge the United States economy into dire straits that would make the 1930s depression seem like an economic boom. The United States may not recover from such a crisis, which would thrust China into the position of the world's leading superpower. Although Scotty was appointed by the president, and technically, the president is everyone's boss, there was still a chain of command everyone had to follow before any information reached the president or his vice president. Scotty's immediate boss was the Secretary of Health and Human Services, Latonya Robinson.

When Latonya received the call from Scotty, she listened intently without saying anything until he was finished. Latonya knew what was at stake as she asked, "Are you absolutely sure you'll have a vaccine soon, and are you certain it will work?"

"I believe we'll have a vaccine soon, but not soon enough, and no one can guarantee any vaccine will work until it's applied," Scotty answered.

There was a sick feeling in Latonya's stomach. "Continue to work on the vaccine as expeditiously as possible. I want you to call me if you run into any bureaucratic red tape."

"I most definitely will," Scotty assured her.

Latonya dreaded the call she knew she would have to make. Her boss was the Vice President of the United States, Earl Roberts.

The White House switch board was a highly classified system for how phone calls coming into the White House were handled. All calls were automatically routed to a secure line that scrambled sound waves into a steady current of electricity, which was the genius of the telephone invented by Alexander Bell in 1876.

After clearing the switch board, Latonya heard the voice of the Vice President of the United States. “Madame Secretary, how can I be of assistance to you?”

“I’m not sure, sir. We have an urgent problem.” Latonya repeated the dire warning Scotty had given her. When she finished, there was silence at the end of the line. “Sir?”

“Yes. I’m here,” the vice president said. “I was digesting what you were telling me and thinking about our options.”

Latonya could hear the resignation in his voice of foreboding doom and thought about the last meeting he and the president walked out on.

Chapter 14

In the still early morning hours, nearing the end of the summer solstice, before the wind would start to pick up just as the sun was starting to rise over the horizon, Skip could barely hear the faint sound of a plane engine.

Derrick Berman and Cab Helms were the pilots and copilots of Striker. It was the latest model in the series of F-16 bomber planes. Lenny Franks was the plane's aerial spotter. "We're over the target, and the sky is clear," he said, looking through the floor telescope.

"Confirmation for the drop, sir," Derrick said into his headset.

"Affirmative," the response came back, sharp and decisive.

The bay doors underneath the plane slowly opened, and along the oblong-shaped object existed the plane descending downwards as the plane climbed steadily upwards and away at a 45-degree angle.

Skip was on his third cup of coffee, thinking about his next plan of action. It had been four days since he had called Bradley Cosgrove. He would have to make a decision soon about what he knew to give the people enough time to make their own decisions and preparations about their lives. Suddenly, all of his senses became heightened, and Skip found himself struggling to breathe. The cup of coffee he was holding slid from his fingers and crashed to the floor, shattering into pieces. At first, Skip thought he was having a heart attack until he looked around the cafeteria. Everyone was grasping for air. Skip tried to understand what was happening, but his thoughts were discombobulated.

The oblong-shaped object was a neutron bomb. It exploded five hundred feet above ground level with a terrifying sound as all the oxygen was sucked out of the air within a hundred-mile radius of the research station. Every living organism that was dependent on oxygen to survive perished in minutes. The only things that were left were buildings, unanimated objects, and the ensuing freezing cold that crept forward like a pestilence. The air deprived of oxygen and warmth that it brings from the sun froze the Antarctica, the Caribbean Sea to the north, the Atlantic Ocean to the east, the Drake Strait of the southern ocean, and the Pacific Ocean to the west. Alaska and large portions of Russia that bordered the Arctic Ocean were also impacted, along with maritime life. No living organism survived the blast. It was a total catastrophe of biblical proportions.

The End

Epilogue

The single most important invention that drove the industrial revolution was the steam engine, developed by Thomas Newcomen and first used in Great Britain in 1712. No one knew the destruction it would leave in its wake.

Fast forward three hundred and eight years later, and scientists now regard the warming of the Earth's climate system as the greatest threat to human existence. Scientists believe this warming is due in large part to human activities and the injection of carbon dioxide into the atmosphere by the burning of fossil fuels that has accompanied industrialization.

Industrialization affects every facet of business and the resulting comforts it brings to humankind in the struggle for survival.

Frankenstein's monster became a monster because its creator lost control of him. Global warming is the new monster of the twenty-first century. Just as the town's people rallied together to destroy Frankenstein's monster, we must—we have to—rally together to harness the political will to address the issue of global warming for our own survival and that of our future generations.

About the Author

Richard Thompson is one of the statistics where one in every four Black men will find themselves enmeshed in America's judiciary system. After languishing for decades—forty-eight years—in various federal and state prisons, he was finally paroled in 2021.

www.ingramcontent.com/pod-product-compliance
Lightning Source LLC
Chambersburg PA
CBHW030849180526
45163CB00004B/1507
9781637847237